11-40

MEI structured mathematics

Numerical Analysis

ELIZABETH WEST

Series Editor: Roger Porkess

MEI Structure Mathematics is supported by industry:

BNFL, Casio, GEC, Intercity, JCB, Lucas, The National Grid Company, Texas Instruments, Thorn EMI

Hodder & Stoughton

LONDON SYDNEY AUCKLAND

MEI Structured Mathematics

Mathematics is not only a beautiful and exciting subject in its own right but also one that underpins many other branches of learning. It is consequently fundamental to the success of a modern economy.

MEI Structured Mathematics is designed to increase substantially the number of people taking the subject post-GCSE, by making it accessible, interesting and relevant to a wide range of students.

It is a credit accumulation scheme based on 45 hour components which may be taken individually or aggregated to give:

3 components	AS Mathematics
6 components	A Level Mathematics
9 components	A Level Mathematics + AS Further Mathematics
12 components	A Level Mathematics + A Level Further Mathematics

Components may alternatively be combined to give other A or AS certifications (in Statistics, for example) or they may be used to obtain credit towards other types of qualification.

The course is examined by the Oxford and Cambridge Schools Examination Board, with examinations held in January and June each year.

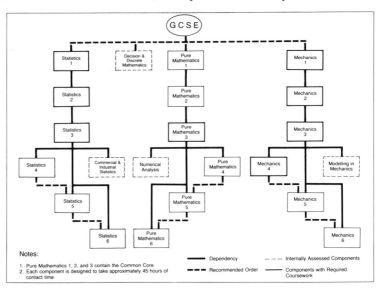

Notes:
1. Pure Mathematics 1, 2, and 3 contain the Common Core.
2. Each component is designed to take approximately 45 hours of contact time.

This is one of the series of books written to support the course. Its position within the whole scheme can be seen in the diagram above.

Mathematics in Education and Industry is a curriculum development body which aims to promote the links between Education and Industry in Mathematics at secondary school level, and to produce relevant examination and teaching syllabuses and support material. Since its foundation in the 1960s, MEI has provided syllabuses for GCSE (or O Level), Additional Mathematics and A Level.

For more information about MEI Structured Mathematics or other syllabuses and materials, write to MEI Office, Monkton Combe, Bath BA2 7HG.

Contents

1 The Solution of Equations   1

2 Approximating Functions   30

3 Numerical Integration   55

4 The Solution of Differential Equations   78

5 Errors in Numerical Processes   100

Answers to Selected Exercises   120

Introduction

This book has been written to meet the needs of students following the Numerical Analysis Component of MEI Structured Mathematics. Numerical analysis is the branch of pure mathematics concerned with obtaining numerical solutions. The text shows how mathematical concepts are used to derive efficient numerical algorithms to find good approximations to the solution of mathematical models of physical problems.

The first four chapters introduce some important topics, namely, the solution of non-linear equations, function approximation, numerical differentiation and integration, and the solution of differential equations. In these chapters, the problems which can arise from approximations of both functions and numerical values become apparent. The final chapter reviews and discusses numerical errors. You may choose to read parts of the final chapter as you work through the rest of the text.

It is assumed that you will have access to a computer or a programmable graphics calculator. The extent to which these are used is left up to you. If you have a computing background you will benefit from programming some of the algorithms; if not there is excellent mathematical software available. You may decide to use a spreadsheet program for most algorithms described in the text.

Each chapter includes a number of routine exercises and it is anticipated that you will work through these as you go through the text. Some of these are extended exercises which are intended to enhance your understanding of the topic. At the end of each chapter is a collection of exercises which give you practice in handling the concepts and algorithms described in the chapter and also some extended questions which could form the basis of coursework tasks.

It is hoped that in working through the book, preferably as one of a group, your curiosity will be aroused, your ability to analyse and solve numerical problems will improve and that you will find the experience rewarding and satisfying.

The author would like to thank Neil Sheldon and Ray Dunnett who helped in the early preparation of this book.

Elizabeth West

1

The Solution of Equations

Many problems which arise in physics, engineering and management science are solved by producing a *mathematical model* of the physical situation. For example, if a ball is dropped from the top of a cliff into the sea 20 metres below, the time taken to fall is given by the solution of the equation

$$s = ut + \frac{1}{2}at^2$$

with $u = 0$, $a = 10\,\mathrm{ms}^{-2}$ and $s = 20$ m. Therefore $20 = 5t^2$ and from this we can deduce that the ball takes 2 seconds to reach the sea. The equation does not describe the physical situation exactly since air resistance has not been taken into account but a time of 2 seconds will be a good approximation to the solution.

Usually, the solution of a mathematical model requires the solution of an equation. Some equations, like

$$2x^2 - 3x - 4 = 0$$

or
$$\cos 2x + \cos x = 0.5$$

can be solved using standard methods already studied. However, in most cases, the equation is less simple. To illustrate the kind of equations which can arise from an everyday problem, consider the following example about calibrating a dipstick.

A cylindrical oil tank is mounted as shown in figure 1.1. The tank has a hole at the top to allow a dipstick to be inserted to measure the volume of oil in the tank. For convenience, the dipstick is calibrated in cubic metres. Given that the tank is 3 m long and 2 m in diameter, where should the marks be placed on the dipstick to indicate 1, 2, 3, 4, . . . m³ of oil in the tank?

As a first step, we calculate the volume of the tank. This is given by

$$\text{(area of cross-section)} \times \text{length} = \pi \times 1^2 \times 3$$
$$= 3\pi \approx 9.42\,\mathrm{m}^3$$

Hence the dipstick should be marked to indicate levels corresponding to 1, 2, 3, . . . 9 m³ of oil in the tank. The problem now is how to determine the depth of oil which corresponds to these volumes.

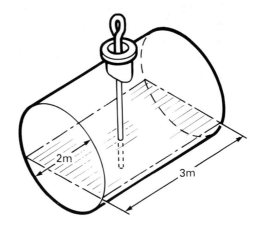

Figure 1.1

Figure 1.2 shows a cross-section of the tank with the oil level AB at a depth of h m.

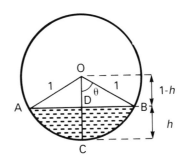

Figure 1.2

Let $\angle COB = \theta$ and note that

$$DB^2 = OB^2 - OD^2$$
$$= 1 - (1 - h)^2$$
$$= 2h - h^2$$
$$\Rightarrow DB = \sqrt{2h - h^2}$$

The volume, V, of oil in the tank is given by

$$V = \text{(length of tank)} \times \text{(cross-sectional area of oil)}$$
$$= 3 \times \text{(area of sector OAB - area of triangle OAB)}$$

$$= 3\left(\frac{1}{2} \times 1^2 \times 2\theta - DB \times (1 - h)\right)$$

$$= 3(\theta - (1 - h)\sqrt{2h - h^2})$$

But $\cos \theta = 1 - h$, so θ is the angle whose cosine is $(1 - h)$; the value of θ is written as arc $\cos(1 - h)$, the function being given by the $\cos^{-1}$ key on your calculator. Therefore we have found that, for levels below half full,

$$V = 3(\text{arc } \cos(1 - h) - (1 - h)\sqrt{2h - h^2})$$

It is not difficult to check that the same formula gives the volume of oil in the case where the tank is more than half full as shown in figure 1.3.

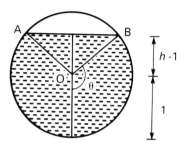

Figure 1.3

To determine the positions of the marks on the dipstick, we must solve the nine equations

$$3 \text{ arc } \cos(1 - h) - 3(1 - h)\sqrt{2h - h^2} = V$$

for $V = 1, 2, 3, \ldots 9$.

Taking the first mark, $V = 1$, the equation becomes

$$3 \text{ arc } \cos(1 - h) - 3(1 - h)\sqrt{2h - h^2} - 1 = 0$$

or $f(h) = 0$

where $\qquad f(h) = 3 \text{ arc } \cos(1 - h) - 3(1 - h)\sqrt{2h - h^2} - 1$

It is, in fact, *not* possible to solve this equation analytically but it is quite possible that this equation has several solutions and we know that the solution we want lies between $h = 0$ and $h = 2$. If we have graph plotting programs and graphics calculators available it is not difficult to obtain the graph of the function f shown in figure 1.4. We see that the root is close to $h = 0.3$, which seems a reasonable answer. In fact, no matter how complicated an equation we are asked to solve, we can determine the approximate values of the roots using a graph plotter.

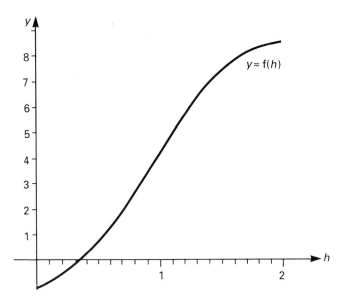

Figure 1.4

Q Each of the following equations has one or more roots in the interval $-5 \le x \le 5$. Use a graph plotter to determine the number of roots in the interval and the approximate value of each root.

(i) $x^3 - 0.5x^2 - 0.5x - 1.5 = 0$
(ii) $1.5x \cos x - \sin x = 0$
(iii) $\sin x = x^2 - 1$.

We found that the dipstick equation
$$3 \arccos(1 - h) - 3(1 - h)\sqrt{2h - h^2} - 1 = 0$$
has a root near to $h = 0.3$. How could we obtain a better approximation? One possibility is to plot the graph over a smaller interval of h values, for example, for $0.25 \le h \le 0.4$, as shown in figure 1.5. The graph shows that the root lies between 0.32 and 0.33. If you have access to a graphics calculator, a similar result could be obtained by zooming in with the cursor.

The disadvantage of this approach is that it is time consuming: Remember that to solve the dipstick problem we still have another 8 equations to tackle. This is the kind of procedure which it should be possible to automate. We would like to find a method which can be programmed to give the solution to an equation as accurately as required and as economically as possible.

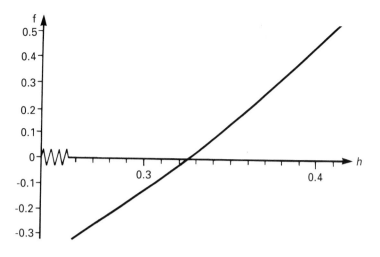

Figure 1.5

The Bisection Method

One possible approach is to note that since the root lies between 0.32 and 0.33, the root is given, approximately, by 0.325; we also know that the error in this approximation is no more than ±0.005.

Taking this further, since

$$f(h) = 3 \arccos(1 - h) - 3(1 - h)\sqrt{2h - h^2} - 1,$$

and $f(0.32) < 0$, $f(0.33) > 0$ we can calculate $f(0.325)$ or use a graph to show that $f(0.325) < 0$.

We can therefore see (figure 1.6) that the graph must cross the axis between $h = 0.325$ and $h = 0.33$. We have narrowed the interval within which we know that the root lies and the mid-point of this interval gives a better approximation as 0.3275.

Clearly the sign of f at $h = 0.3275$ could be calculated and a further narrowing of the interval carried out. This *bisection* of the interval

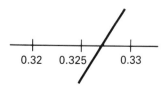

Figure 1.6

containing the root can be repeated until the root is obtained to any required level of accuracy. All that is required at each step is the calculation of the mid-point of an interval, the value of f at that point and a decision as to which of the two half intervals contains the root. Each of these operations can be programmed easily.

When computation of this type is being done by hand, it is important that the work is set down clearly so that it can be checked easily. One way of setting down the calculation is in a table of the following form.

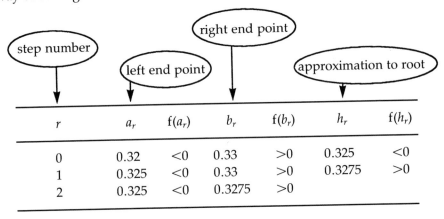

r	a_r	$f(a_r)$	b_r	$f(b_r)$	h_r	$f(h_r)$
0	0.32	<0	0.33	>0	0.325	<0
1	0.325	<0	0.33	>0	0.3275	>0
2	0.325	<0	0.3275	>0		

Q Complete step 2 in the above calculation to obtain a more accurate approximation to the solution of the equation

$$3 \arccos(1-h) - 3(1-h)\sqrt{2h-h^2} - 1 = 0.$$

This is an example of an *iterative process* where the procedure for repeatedly halving the interval width is continued until the required accuracy is obtained. This method is called the *Bisection Method*.

For a given equation $f(x) = 0$ with a root lying between $x = a_0$ and $x = b_0$, the method can be described using a flowchart (figure 1.7).

For a numerical algorithm to be useful, it must be reliable. Note that it is not enough to check that $f(a_0)$ and $f(b_0)$ are of different signs. For example, for the equation

$$\tan x = 0$$

since $\tan 1.5 > 0$, and $\tan 1.6 < 0$, we might expect a root to lie in the interval $(1.5, 1.6)$ and start to use the algorithm. Examine the graph of $y = \tan x$ and explain why the method would *not* give a root in this case.

However, for any root of an equation $f(x) = 0$, it will always be possible to find an interval (a_0, b_0) which contains the root and for which the bisection method will give the root to any required degree of accuracy.

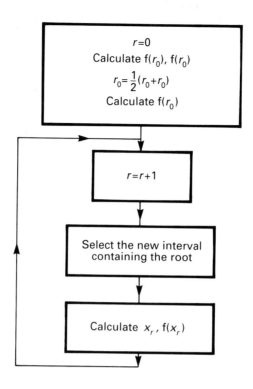

Figure 1.7

EXAMPLE   Use a graph plotting program to verify that the equation

$$x^3 + 4x^2 - 10 = 0$$

has a root in the interval $1 \leq x \leq 2$.

Use the bisection method to obtain this root correct to 1 decimal place.

Solution:
Let $f(x) = x^3 + 4x^2 - 10$.

The graph of the function f is shown in figure 1.8 and the equation $f(x) = 0$ has a root in the interval $1 \leq x \leq 2$.

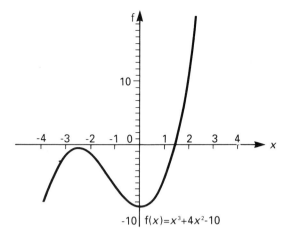

Figure 1.8

Using the bisection method, we obtain the following table.

r	a_r	$f(a_r)$	b_r	$f(b_r)$	x_r	$f(x_r)$
0	1	<0	2	>0	1.5	>0
1	1	<0	1.5	>0	1.25	<0
2	1.25	<0	1.5	>0	1.375	>0
3	1.25	<0	1.375	>0	1.3125	<0
4	1.3125	<0	1.375	>0	1.34375	<0
5	1.34375	<0	1.375	>0	1.359375	<0
6	1.359375	<0	1.375	>0		

At step 6, we see that the root lies between 1.359375 and 1.375, so, on rounding to 1 decimal place, the root must be 1.4.

Q

1. Each of the equations

    (i) $x^3 - 0.5x^2 - 0.5x - 1.5 = 0$

    and (ii) $1.5x \cos x - \sin x = 0$

    was shown to have a root near to $x = 1$. Use the bisection method to determine each of these roots correct to 1 decimal place.

2. Continue the solution of the equation arising from the dipstick problem, namely,

$$3 \arccos(1 - h) - 3(1 - h)\sqrt{2h - h^2} - 1 = 0$$

    to obtain the root correct to 3 decimal places.

Investigation

Investigate the number of iterations of the bisection method which are required to find a root to a given accuracy as follows.

1. We saw earlier that the equation

$$x^2 - \sin x - 1 = 0$$

has a root between $x = 1$ and $x = 2$.

Let $\varepsilon_r = x_r - a_r = $ the magnitude of the maximum possible error at the rth step. Copy the following table and complete it by carrying out 4 steps of the bisection method starting with $a_0 = 1$ and $b_0 = 2$.

r	a_r	$f(a_r)$	b_r	$f(b_r)$	x_r	$f(x_r)$	ε_r
0	1	-0.8415	2	2.0907	1.5	0.2525	0.5
1							
2							
3							
4							

2. Examine the values in the last column of the table. What do you notice about these values?

   Write down the magnitude of the maximum possible error after

   (i) 5 steps   (ii) 10 steps   (iii) p steps.

3. How many iterations must be carried out so that the error is reduced to less than 0.00001 in magnitude?

From the investigation carried out above, we realise that, *irrespective of what equation we are solving*, if the starting interval is of width 1 unit, it will take 16 steps to reduce the error in the estimate to less than 0.00001 in magnitude. Indeed, for any starting interval (a_0, b_0) the errors at each step are given by:

Step no	0	1	2	$\ldots$	p
Max. error	$\dfrac{1}{2}(b_0 - a_0)$	$\dfrac{1}{2^2}(b_0 - a_0)$	$\dfrac{1}{2^3}(b_0 - a_0)$	$\ldots$	$\dfrac{1}{2^{p+1}}(b_0 - a_0)$

Suppose that we start with an interval of width 1 unit and we require the root correct to 3 decimal places.

NA

Then the error in the final step should be less than 0.5×10^{-3}

so that $$\left(\frac{1}{2}\right)^{p+1} < 0.5 \times 10^{-3}$$

Taking logarithms, $(p+1)\log(0.5) < \log(0.0005)$

$$\Rightarrow \quad p+1 > \frac{3.301}{0.301}$$

$$\Rightarrow \quad p > 9.97$$

showing that at least 10 iterations will be required.

> **Q** | How many iterations would you expect to have to carry out to solve the equations
>
> (i) $e^x - 3x = 0$, $\qquad a_0 = 0, b_0 = 1$
> and (ii) $x^3 - 3x + 1 = 0$, $\qquad a_0 = 1, b_0 = 2$
>
> correct to 2 decimal places, using the starting intervals given?
>
> Carry out the computation and note how many steps were required in each case. Comment on your results.

It would not be a difficult task to write a simple program to carry out the steps of the bisection method. Using such a program, with suitable starting values a_0 and b_0, the 9 equations arising from the dipstick problem could be solved and the task completed. The bisection method is a good, reliable method for solving an equation $f(x) = 0$ but it might be worth investigating whether or not a more efficient method could be found.

Fixed Point Iteration

Experiment

Key any positive number into your calculator and obtain its square root; now obtain the square root of the number displayed and continue in this way for some time. What happens?

Now, set your calculator to radian mode. Key in any number in the interval $-10 \le x \le 10$ and obtain the cosine repeatedly, in the same way.

The above process could be described by the diagram in figure 1.9.

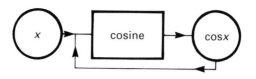

Figure 1.9

You should find that, in the first case the value displayed becomes closer and closer to 1 and in the second case, to 0.739085133. What is the significance of these numbers?

Suppose that we denote the number entered in the calculator as x_0 then, pressing the square root key gives a value x_1 with

$$x_1 = +\sqrt{x_0}$$

The process is repeated, generating a sequence of numbers $x_0, x_1, x_2 \ldots .$. This sequence can be described by the formula

$$x_{r+1} = \sqrt{x_r} \text{ with } x_0 \text{ given, } \text{ for } r = 0, 1, 2, \ldots$$

with each member being calculated by taking the square root of the previous member. A formula of this type is called a *recurrence relation* and we found in the calculator experiment that $x_r \to 1$ as $r \to \infty$.

Similarly, when the cosine key is used, we generate values of x_r where

$$x_{r+1} = \cos x_r, \quad x_0 \text{ given, } \text{ for } r = 0, 1, 2, \ldots .$$

In this case, we find that

$$x_r \to 0.739085133 \quad \text{as } r \to \infty.$$

This does not happen with all the function keys on a calculator for example, try using the 'tan' key in place of the 'cos' key in the above experiment.

In some cases, the values of x_r tend to a limit as r approaches ∞ and the sequence $x_0, x_1, x_2 \ldots$ so formed is said to be a *convergent* sequence. In the example using the square root function, if pressing the square root key makes no difference to the value displayed, we know that x is equal (or approximately equal) to its own square root, so that

$$x = \sqrt{x}$$
$$\Rightarrow \quad x^2 = x$$
$$\Rightarrow \quad x = 0 \text{ or } x = 1.$$

So the sequence converged to one of the roots of the equation $x = \sqrt{x}$. Similarly, $x = 0.739085133$ is a root of the equation $x = \cos x$.

Extending this approach, for *any* equation $f(x) = 0$, it is possible to rearrange the equation to the form $x = g(x)$. It may be that the recurrence relation

$$x_{r+1} = g(x_r), \quad x_0 \text{ given},$$

produces a convergent sequence, and, *if* this is the case, the limit of the sequence will be a root of the equation $f(x) = 0$.

To illustrate this, consider the equation

$$x^2 - \sin x - 1 = 0$$

which we found to have a root between $x = 1$ and $x = 2$. This equation could be rearranged in various ways, for example,

(i) $x = \sqrt{1 + \sin x}$
(ii) $x = \arcsin(x^2 - 1)$
(iii) $x = x - (x^2 - \sin x - 1)$

(iv) $x = x - \dfrac{1}{2}(x^2 - \sin x - 1)$

Could any of these rearrangements be used to solve the equation? Using each of these to give a recurrence relation $x_{r+1} = g(x_r)$ with $x_0 = 1$, gives the following sequences in which the values shown have been rounded to 4 decimal places.

r	(i)	(ii)	(iii)	(iv)
0	1	1	1	1
1	1.3570	0	1.8415	1.4207
2	1.4061	-1.5708	0.4140	1.4059
3	1.4094	–	1.6449	1.4108
4	1.4096	–	0.9364	1.4092
5	1.4096	–	1.8650	1.4098

Note that the sequences shown in columns (i) and (iv) appear to be converging to a value near to 1.410. The sequence in column (iii) is oscillating and does not appear to be converging. The sequence $x_{r+1} = \arcsin(x_r^2 - 1)$, $x_0 = 1$, whose values are listed in column (ii), gives $x_2 = -1.5708$ and since $x_2^2 - 1 > 1$, the value of $\arcsin(x_2^2 - 1)$ is not defined.

Using 10 steps of the bisection method with $a_0 = 1$ and $b_0 = 2$ gives the root as lying between 1.4082 and 1.4102. It looks as though we have found a method which, *when* it works, converges much faster than the bisection method.

Any solution of the equation $x = g(x)$ is called a *fixed point* of the function g. It can be shown that, if a function g has a fixed point and if the iterative sequence defined by

$$x_{r+1} = g(x_r), \quad x_0 \text{ given,}$$

converges then the sequence converges to a fixed point of g. For this reason, the method of solving an equation $f(x) = 0$ by rearranging it to the form $x = g(x)$ and generating an iterative sequence, $x_{r+1} = g(x_r)$, with (x_0) given is called *Fixed Point Iteration*.

If this method is to be of any use, we must investigate under what circumstances the sequence converges and what governs the rate at which it converges.

Investigation

Investigate the convergence of the following fixed point iteration.

1. Using a graph plotter, sketch the graphs of $y = x$ and $y = 1.2 - \cos(2x - 0.8)$ for $-1 \le x \le 5$.

   Write down the approximate value of the fixed points of the function g where

   $$g(x) = 1.2 - \cos(2x - 0.8)$$

   Explain why each fixed point is a root of the equation

   $$x + \cos(2x - 0.8) = 1.2$$

2. By writing a simple program to do the computation, or otherwise, investigate the behaviour of the sequence generated by

   $$x_{r+1} = 1.2 - \cos(2x_r - 0.8)$$

   with $x_0 = 0, 0.5, 1, 1.1, 1.2$ and 1.5.

   You might find it interesting to experiment with other starting values. Comment on the results you obtain.

EXAMPLE

Show that, if the following two sequences converge as $r \to \infty$, they will converge to either $x = -1$ or $x = 2$.

(i) $x_{r+1} = x_r^2 - 2$, x_0 given
(ii) $x_{r+1} = \sqrt{2 + x_r}$, x_0 given.

Using starting values of 0 and 3, calculate the first five members of each sequence. Comment on the values obtained.

Solution:
For sequence (i), if $x_r \to x$ as $r \to \infty$,

$$x = x^2 - 2$$

$$\Rightarrow \quad x^2 - x - 2 = 0$$
$$\Rightarrow \quad (x - 2)(x + 1) = 0$$
$$\Rightarrow \quad x = 2 \text{ or } -1.$$

Similarly for sequence (ii), if $x_r \to x$, as $r \to \infty$

$$x = \sqrt{2 + x}$$
$$\Rightarrow \quad x^2 = 2 + x$$
$$\Rightarrow \quad x^2 - x - 2 = 0$$
$$\Rightarrow \quad x = 2 \text{ or } -1.$$

Calculating the first 5 members of the sequences, and showing the values rounded to 4 significant figures gives the following table:

	$x_{r+1} = x_r^2 - 2$		$x_{r+1} = \sqrt{2 + x_r}$	
r	$x_0 = 0$	$x_0 = 3$	$x_0 = 0$	$x_0 = 3$
0	0	3	0	3
1	-2	7	1.414	2.236
2	2	47	1.848	2.058
3	2	2209	1.962	2.014
4	2	4.880×10^6	1.990	2.003
5	2	2.381×10^{13}	1.997	2.007

In three of the four sequences examined, convergence to $x = 2$ takes place. One *diverges*, successive values becoming larger and larger.

The Criterion for Convergence in Fixed Point Iteration

It often helps our understanding of a mathematical process if we can represent the process graphically. Consider again the iterative formula studied earlier namely

$$x_{r+1} = 1.2 - \cos(2x_r - 0.8)$$

The graphs of $y = x$ and $y = 1.2 - \cos(2x - 0.8)$ are shown in figure 1.10.

At fixed points of a function g, $x = g(x)$ so that the points P, Q and R in figure 1.10 are fixed points of the function

$$g(x) = 1.2 - \cos(2x - 0.8)$$

Figure 1.11 shows an enlarged section of figure 1.10 near to P and taking $x_0 = 0.5$, we know that $x_1 = 1.2 - \cos(2x_0 - 0.8)$.

The value, x_1, is represented by the length of the line AB in the diagram and since C lies on the line $y = x$,

$$x_1 = y_B = y_C = x_C$$

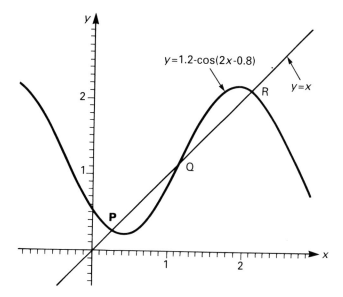

Figure 1.10

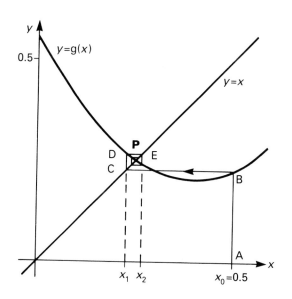

Figure 1.11

This process can be repeated giving

$$x_2 = y_D = y_E = x_E$$

Continuing in this way we produce a 'cobweb' which centres on the fixed point of the function at P. The steps taken to form a cobweb diagram are:

(i)   draw the line parallel to the y-axis through $x = x_0$, to meet the curve $y = g(x)$ at B;
(ii) draw the line through B parallel to the x-axis to meet $y = x$ at C;
(iii) draw the line through C parallel to the y-axis to meet $y = g(x)$ at D;
(iv) draw the line through D parallel to the x-axis to meet $y = x$ at E; etc.

Q | On a graph of $y = x$ and $y = 1.2 - \cos(2x - 0.8)$, construct the cobweb diagram obtained from the iterative process

$$x_{r+1} = 1.2 - \cos(2x_r - 0.8) \, , \ x_0 = 1.5$$

This shows that, with a starting value of $x_0 = 1.5$, the iterative process

$$x_{r+1} = 1.2 - \cos(2x_r - 0.8)$$

converges to the fixed point at R. If we experiment with different starting values in the interval $0.5 \leq x_0 \leq 2$, we obtain the results shown in figure 1.12. We find that, for values of $x_0 < x_Q$, the sequence converges to the value x_P and for $x_0 > x_Q$, the values converge to x_R.

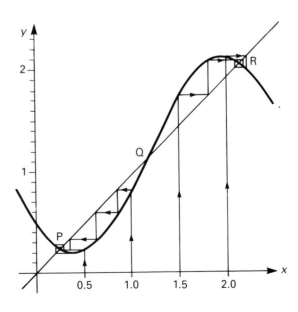

Figure 1.12

As a second illustration, consider the iterative formula

$$x_{r+1} = 0.2e^{x_r} + 0.2$$

As before, we sketch graphs of $y = x$ and $y = 0.2e^x + 0.2$, as shown in figure 1.13; this indicates that if $g(x) = 0.2e^x + 0.2$, the function g has two fixed points, one at P, the other at Q.

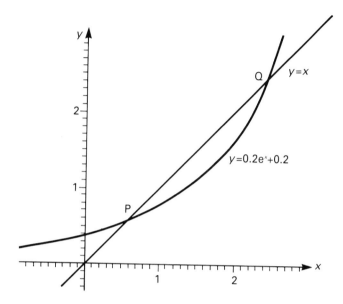

Figure 1.13

Taking $x_0 = 1$ and proceeding as before by drawing a line up to the curve $y = 0.2e^x + 0.2$ and across to the line $y = x$ and so on, we can produce a 'staircase' diagram which gives successive x_r values closer and closer to the fixed point, P, near to $x = 0.5$. Again, taking different values of x_0 in the interval $1 \leq x_0 \leq 2.5$ gives the results shown in figure 1.14.

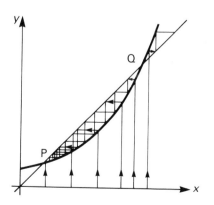

Figure 1.14

We see that, when convergence takes place, the values converge towards the fixed point, P. For values of $x_0 > x_Q$ the sequence values diverge to infinity.

Note that, in a cobweb diagram, the sequence values *oscillate* about the fixed point but in a *staircase* diagram the values are all on one side of the fixed point. We find that if the slope of the curve $y = g(x)$ is positive at the fixed point, a staircase diagram is obtained near the fixed point and if the slope is negative, the diagram has cobweb form. We must continue to investigate the factor which determines whether or not convergence will take place.

If you look closely at the two examples used in the last section, you may notice that the sequence failed to converge to the fixed point nearest to x_0 in the cases where the slope of the curve $y = g(x)$ at that fixed point was steep, either sloping steeply upwards or steeply downwards. By analysing the iterative process we can show that the gradient of the curve $y = g(x)$ at a fixed point is the factor which determines whether or not convergence to that point will take place.

Suppose that a sequence $x_0, x_1, x_2 \ldots$ is obtained from the iterative formula $x_{r+1} = g(x_r)$, with x_0 given and that the function g has a fixed point at $x = a$.

Let
$$x_r = a + \varepsilon_r$$

so that ε_r can be considered as the 'error' in the rth iterate.

Then, if the sequence converges to a, $\varepsilon_r \to 0$ as $r \to \infty$.

But
$$x_{r+1} = g(x_r)$$

so that
$$a + \varepsilon_{r+1} = g(a + \varepsilon_r) \tag{1}$$

From figure 1.15
$$g(a + \varepsilon_r) = PQ + QR = g(a) + QR$$

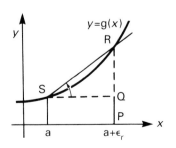

Figure 1.15

If ε_r is small,

$$\frac{QR}{SQ} = \tan \angle QSR$$

$$\approx \text{ slope of } y = g(x) \text{ at } x = a$$

$$\approx g'(a)$$

Therefore $\qquad QR \approx \varepsilon_r g'(a)$ and from (1),

$$a + \varepsilon_{r+1} = g(a + \varepsilon_r)$$

$$\approx g(a) + \varepsilon_r g'(a)$$

But, since a is a fixed point of g, $a = g(a)$ and hence $\varepsilon_{r+1} \approx \varepsilon_r g'(a)$.

So, the error at the $(r + 1)^{\text{th}}$ step is given, approximately, by the product of $g'(a)$ and the error at the rth step.

Since we are concerned with the *magnitude* of the error, we note that

$$|\varepsilon_{r+1}| \approx |g'(a)| . |\varepsilon_r| \qquad (2)$$

and hence $\qquad |\varepsilon_{r+1}| < |\varepsilon_r|$ if $|g'(a)| < 1$.

Therefore, if $|g'(a)| < 1$, the magnitude of the error in successive values of x_r will decrease and x_r will approach a as $r \to \infty$.

This result can be stated more formally, as follows:

If the function g has a fixed point at $x = a$ and if g' exists near to $x = a$ then

- if $|g'(a)| < 1$ and a suitable starting-point is chosen, the sequence defined by $x_{r+1} = g(x_r)$ will converge to $x = a$
- the starting-value x_0 will be a suitable starting-value if x_0 lies within the interval containing a for which $|g'(x)| < 1$.

In practice, if x_0 is a first approximation to the fixed point, the iterative scheme will probably converge if $|g'(x_0)| < 1$.

It is important to note that, from (2),

$$|\varepsilon_{r+1}| \approx |g'(a)| \varepsilon_r$$

and so the closer $|g'(a)|$ is to zero, the faster will be the convergence.

EXAMPLE

Use a graph plotter to verify that the equation

$$0.5 \sin x - x + 3 = 0$$

has a root in the interval $0 \le x \le 2\pi$. Choose a suitable iterative formula to locate this root correct to 2 decimal places.

Illustrate the convergence on a cobweb/staircase diagram.

Solution:

The graph of $y = 0.5 \sin x - x + 3$ is shown in figure 1.16, indicating that the equation $0.5 \sin x - x + 3 = 0$ has a root near to $x = 3$.

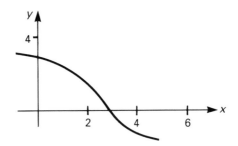

Figure 1.16

To find a suitable iterative formula, try

$$x = 0.5 \sin x + 3$$

so that

$$g(x) = 0.5 \sin x + 3$$

and

$$g'(x) = 0.5 \cos x$$

Since $-1 \leqslant \cos x \leqslant 1$ for all x, $|g'(x)| \leqslant 0.5$ and so the sequence

$x_{r+1} = g(x_r) = 0.5 \sin x_r + 3$, $x_0 = 2.5$, will converge.

The sequence of values generated is shown in the following table.

r	x_r
0	2.5
1	3.29924
2	2.92150
3	3.10916
4	3.01621
5	3.06252
6	3.03949
7	3.05096
8	3.04525

The values of x_r are oscillating and, correct to 2 decimal places, the root is 3.05. The cobweb diagram is shown in figure 1.17, taking $x_0 \approx 2.5$.

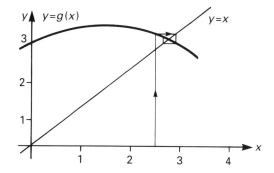

Figure 1.17

Q

Verify that each of the equations

  (i) $x - \cos x = 0$

and (ii) $3x^2 - e^x = 0$

has a root which lies between $x = 0$ and $x = 1$. Determine an iterative formula which could be used to give the root, in each case.

Obtain each root correct to 2 decimal places.

The serious disadvantage of the method of fixed point iteration for solving $f(x) = 0$ lies in the process of determining an iterative formula which will lead to fast convergence. Essentially it is a trial and error process which cannot be automated easily. However, the idea of fixed point iteration underlies the following method in which the iterative formula can be derived from the function f.

The Newton-Raphson Method

This method can be derived most easily from geometrical considerations. Suppose that the equation $f(x) = 0$ has a root at $x = a$, and that x_0 is known to be close to a as shown in figure 1.18.

Let P be the point with coordinates $(x_0, f(x_0))$ and let the tangent at P meet the x-axis at the point Q at which $x = x_1$. In general, x_1 will be a better approximation to the root. The process can be repeated to give a sequence of points $x_2, x_3 \ldots$ each one closer to the value a, until the desired accuracy is obtained.

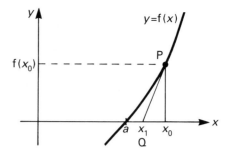

Figure 1.18

To determine an expression for x_1:
Since the slope of the tangent to $y = f(x)$ at $x = x_0$ is $f'(x_0)$, the equation of QP is

$$\frac{y - f(x_0)}{x - x_0} = f'(x_0)$$

$$\Rightarrow \qquad y - f(x_0) = f'(x_0)(x - x_0)$$

At Q, $y = 0$ and $x = x_1$ so that

$$0 - f(x_0) = f'(x_0)(x_1 - x_0)$$

and, if $f'(x_0) \neq 0$, $\qquad x_1 = x_0 - \dfrac{f(x_0)}{f'(x_0)}$

Hence, the iterative formula is

$$x_{r+1} = x_r - \frac{f(x_r)}{f'(x_r)}.$$

This formula, used with a suitable first approximation to the root, is the basis of the *Newton-Raphson method*.

HISTORICAL NOTE

The method was devised by Isaac Newton in the seventeenth century and first published by Joseph Raphson in 1690. Newton's contribution to science is among the greatest of all time; mathematics was only one area in which he worked and his great rival in that field, Leibnitz, said of him 'Taking mathematics from the beginning of the world to the time of Newton, what he has done is much the better half'. He was a stimulus to many in the following century and yet, at the end of his life he stated 'I do not know what I may appear to the world; but to myself I seem to have been only like a boy playing on the seashore, and diverting myself in now and then finding a smoother pebble or a prettier shell than ordinary, whilst the great ocean of truth lay all undiscovered before me'.

1. *The Newton-Raphson formula is a fixed-point iterative formula since, if*

$$g(x) = x - \frac{f(x)}{f'(x)},$$

$x = g(x)$ is a rearrangement of the equation $f(x) = 0$. However, it is not a rearrangement which you would find by chance!

2. *Problems can arise if $f'(x)$ is zero near to the root.*

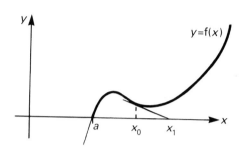

Figure 1.19

In figure 1.19 if x_0 is taken as the first approximation to the root at $x = a$, x_1 is in fact further from the root than x_0. However had a starting value closer to the root been chosen, the method would have converged to $x = a$.

EXAMPLE

By applying the Newton-Raphson method to the equation $x^2 - 7 = 0$, derive an iterative formula for approximating $\sqrt{7}$. Hence obtain $\sqrt{7}$ correct to 3 decimal places.

Solution:

Let $\qquad\qquad\qquad\qquad$ $f(x) = x^2 - 7$

so that $\qquad\qquad\qquad$ $f'(x) = 2x$

Using the Newton-Raphson method,

$$x_{r+1} = x_r - \frac{x_r^2 - 7}{2x_r}, \quad \text{with } x_0 = 3,$$

gives $\qquad\qquad$ $x_1 = 3 - \dfrac{9 - 7}{6} = 2.67$

$$x_2 = 2.67 - \frac{7.1289 - 7}{5.34} = 2.646$$

$$x_3 = 2.646 - 0.0002 = 2.646 \text{ correct to 3 decimal places.}$$

Note that the convergence to 3 decimal places required only 3 iterations.

Q Use the Newton-Raphson method to solve the equations

(i) $x + \cos x = 0$ and   (ii) $x - 3 \sin x = 1$

correct to 4 decimal places, noting how many iterations are required.

The Rate of Convergence of the Newton-Raphson Method

From the above examples, the Newton-Raphson method appears to converge fast. Since it is a fixed point iterative method of the form

$$x_{r+1} = g(x_r)$$

with
$$g(x) = x - \frac{f(x)}{f'(x)}$$

we can examine the rate of convergence by looking at the value of $g'(a)$ where $x = a$ is the root of the equation.

Using the quotient rule for differentiation we have,

$$g'(x) = 1 - \frac{f'(x).f'(x) - f(x).f''(x)}{\{f'(x)\}^2}$$

Hence,
$$g'(a) = 1 - \frac{\{f'(a)\}^2 - f(a)f''(a)}{\{f'(a)\}^2}$$

$$= 1 - 1 \quad \text{since } f(a) = 0$$
$$= 0$$

and since for a fixed point iterative method,

$$|\varepsilon_{r+1}| \approx |g'(a)| \times |\varepsilon_r|$$

the error in the $(r + 1)$ step will be very small.

A more careful analysis which will be carried out in an exercise at the end of Chapter 2 shows that for the Newton-Raphson method

$$\varepsilon_{r+1} \propto \varepsilon_r^2$$

So, as long as x_0 is close to the root at $x = a$, making ε_0 small, the errors will almost always decrease very rapidly.

If an iterative process has the property that

$$\varepsilon_{r+1} \propto \varepsilon_r^2$$

then the iteration is said to be of *second order*; if $\varepsilon_{r+1} \propto \varepsilon_r$, it is of *first order*. As long as certain conditions on the derivatives of the function f are satisfied, each application of the Newton-Raphson formula *approximately* doubles the number of correct decimal places in the approximation. Hence if x_0 is correct to 1 decimal place, x_1 will *probably* be correct to 2 decimal places, x_2 to four, and so on.

When assessing the efficiency of this method, it must be remembered that two functions, namely f(x) and f'(x) must be computed at each step. In some cases, it can be difficult to obtain and evaluate f'(x). In practice, when solving equations by hand, the Newton-Raphson method will probably be best. However the difficulties which can arise from the nature of f'(x) make the slower bisection method more attractive for automatic computation.

Exercise 1

1. Show that the equation $x^3 - x - 1 = 0$ has exactly one root in the interval (1, 2). Use the bisection method to obtain this root correct to 1 decimal place.

2. Show that the equation $x^2 + \ln x = 0$ has only one root. An approximation to the root is required with an error of less than 0.05. Use the bisection method to obtain the approximation.

3. Use the bisection method to obtain the root of the equation $x = \tan x$ which lies in the interval $4 \leqslant x \leqslant 4.5$, correct to 1 decimal place.

4. Find the roots of the following equations, correct to 2 decimal places. You may find it useful to use a computer program or package to assist with the computation.

    (i) $x + \cos x = 0$
    (ii) $3xe^x = 1$
    (iii) $xe^{-x} = 0.25$

5. Show that, if any of the following iterative formulae converge, they will converge to a root of the equation $x^2 - 6x + 5 = 0$.

    (i) $$x_{r+1} = 6 - \frac{5}{x_r}$$

    (ii) $$x_{r+1} = \frac{(x_r^2 + 5)}{6}$$

    (iii) $$x_{r+1} = \sqrt{(6x_r - 5)}$$

    Without carrying out any iterations, determine whether or not any of the formulae do converge to either root, assuming that a suitable starting value is chosen.

6. Determine whether or not the iterative formula

    $$x_{r+1} = x_r^2 - \frac{1}{x_r}$$

    with a suitable starting value, would converge to the root of the equation

    $$x^3 - x^2 - 1 = 0$$

    which is near to 1.5.

7. Show that the sequences defined by

(i)
$$x_{r+1} = \frac{1}{3}(x_r^2 + 2)$$

and (ii)
$$x_{r+1} = 3 - \frac{2}{x_r}$$

with suitable starting values will converge to different roots of the same equation. In each case, construct cobweb or staircase diagrams to show the behaviour near the roots.

8. Prove that the sequence defined by

$$x_{r+1} = \frac{1}{2}\left(x_r + \frac{2}{x_r}\right)$$

converges to $\sqrt{2}$ for any $x_0 > \sqrt{2/3}$.

9. Examine the members of the sequences obtained from

$$x_{r+1} = x_r + \frac{1}{x_r^3}$$

with $x_0 = 1, 4, 5, 10$ and 20.

Let $g(x) = x + \frac{1}{x^3}$ and by examining the graphs of $y = x$ and $y = g(x)$

explain the results you have obtained above.

10. Obtain the value of $\sqrt[3]{25}$ correct to two decimal places using both the bisection method and the Newton-Raphson method. Note the number of iterations required in each case.

11. Use the Newton-Raphson method, with the starting values given, to solve the following equations correct to 3 decimal places.

(i) $\cos x - 0.6x = 0$, $x_0 = 1$
(ii) $2\cos x + 5x - 1 = 0$, $x_0 = 0$
(iii) $x^4 + x - 3 = 0$, $x_0 = 1.5$
(iv) $x^4 - 2 = 0$, $x_0 = 1.5$

12. Show, graphically, that the equation

$$4\sin x = 1 + x$$

has 3 roots. Obtain the roots, correct to 3 significant figures, using the Newton-Raphson method.

13. Determine all the roots of the following two equations, correct to 4 decimal places:

(i) $x^2 - 3x - \frac{1}{(x+1)} = 0$

(ii) $2x - 20\sin x + 1 = 0$, $x \geqslant 0$.

14. Show that the equation

$$x^3 - 5x + 4.2 = 0$$

has two roots in the interval $(1, 1.5)$. Using a computer program to assist with the computation, investigate the result of using the Newton-Raphson method, with $x_0 = 1, 1.1, 1.2, 1.3, 1.4$ and 1.5, to solve the equation. Determine the number of iterations required to give convergence to 3 decimal places in each case. Explain your results with the aid of a diagram.

Determine the negative root of the equation. By making a small change to the constant term in the given equation, obtain an equation with only one real root. An equation for which a small change in the equation gives a significant change in the solution, for example, from 3 real roots to only 1 real root as in this example, is said to be *ill-conditioned*. We will look at ill-conditioned problems in more detail in Chapter 5.

15. In the diagram below, $AD = 20$ units, $BC = 30$ units and $EF = 8$ units. Write down expressions for the lengths of AB and CD in terms of the length, w, of the line AC.

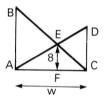

Using similar triangles, or otherwise, obtain an expression for the length of CF in terms of w. Repeat this to obtain an expression for AF in terms of w. Hence write down an equation which must be satisfied by w and solve this equation.

16. Consider the equation $f(x) = 0$ where

$$f(x) = (x - 1)^3(x - 2)$$
$$= x^4 - 5x^3 + 9x^2 - 7x + 2$$

This equation has a root at $x = 2$ and a triple root at $x = 1$. Use a computer program to evaluate the two roots correct to 3 decimal places, using the Newton-Raphson method, using $x_0 = 0.5, 1.5$ and 2.5. Note the number of iterations required for convergence in each case and comment on the results.

Extended Questions

1. The equation

$$e^x \sin 2x - Ax^3 = 0, \qquad (1)$$

where A is a constant with $0 \le A \le 1$, is to be solved for values of x in the interval $0 \le x \le 2\pi$. By examining graphs of $y = e^x \sin 2x$ and $y = Ax^3$, discuss the number and approximate values of the roots of the given equation.

Obtain all the roots of the equation, correct to 4 decimal places, for (i) A = 1/2 (ii) A = 3/4 and (iii) A = 1. Show that if an equation $f(x) = 0$ has a multiple root at $x = a$, the equation $f'(x) = 0$ will also have a root at $x = a$. Hence obtain an equation, independent of A, which is satisfied by the double root of the equation (1), which lies between 0 and 2π. Determine this double root of (1) and the value of A for which it occurs, correct to 5 significant figures.

2. By considering the values of the constant k for which the iterative formula

$$x_{r+1} = x_r + kf(x_r)$$

will converge with a suitable x_0, derive the Newton-Raphson formula. In the Newton-Raphson method, two functions, namely $f(x_r)$ and $f'(x_r)$ are computed at each step. You may have noticed in examples that the value of $f'(x_r)$, $r = 0, 1, 2, \ldots$ does not usually change significantly at each iteration. Would you expect the iterative formula

$$x_{r+1} = x_r - \frac{f(x_r)}{f'(x_0)}$$

with a suitable starting value to give convergence to a root of $f(x) = 0$?

By solving some selected equations using both methods, compare the efficiency of this modified method with that of the Newton-Raphson method.

3. In some cases, it is difficult to obtain and evaluate the function $f'(x)$ for the solution of the equation $f(x) = 0$ using the Newton-Raphson method.

It has been suggested that $f'(x_r)$ could be approximated by

$$f'(x_r) = \frac{f(x_r) - f(x_{r-1})}{x_r - x_{r-1}}.$$

This method, called the *secant method*, requires two starting values x_0 and x_1. Look at the geometrical interpretation of this method. Compare the results obtained from the application of the secant method and the Newton-Raphson method on a selection of equations.

4. For a first order iterative process in which x_{r-1}, x_r and x_{r+1} are successive iterates converging to a fixed point a, it is known that $\varepsilon_{r+1} = k\varepsilon_r$ for some constant k, where $\varepsilon_r = a - x_r$.

By rearranging the equations

$$a - x_{r+1} \approx k(a - x_r)$$

and

$$a - x_r \approx k(a - x_{r-1})$$

obtain an expression for a in terms of x_{r-1}, x_r and x_{r+1}.

This procedure for accelerating the convergence is known as *Aitken's method*. Investigate the efficiency of the method.

KEY POINTS

- The numerical solution of an equation $f(x) = 0$ has two parts:

  (i) the determination of first approximations to the roots

  (ii) the successive improvement of these approximations, to give the roots to any required degree of accuracy.

- If a root lies between $x = x_0$ and $x = x_1$, then

$$x_2 = \frac{1}{2}(x_0 + x_1)$$

  will give an approximation to the root. In the bisection method, this process is repeated using intervals whose widths are successively halved.

- The point $x = a$ is called a fixed point of the function g if $a = g(a)$.

- If $|g'(a)| < 1$ and x_0 is sufficiently close to a, the sequence generated by

$$x_{r+1} = g(x_r)$$

  will converge to the value a.

- If $x_r = a + \varepsilon_r$ and $\varepsilon_{r+1} \propto \varepsilon_r$ then a converging sequence x_0, x_1, x_2 . . . is said to have first order convergence. If $\varepsilon_{r+1} \propto \varepsilon_r^2$, the convergence is of second order.

- The sequence of values generated by

$$x_{r+1} = x_r - \frac{f(x_r)}{f'(x_r)}, \ x_0 \text{ given}$$

  will usually converge to a root of $f(x) = 0$ near to $x = x_0$. The method is called the Newton-Raphson method and has second order convergence.

2 Approximating Functions

Have you ever considered how you might work out the value of the sine or cosine function for a given value of the variable x if you did not have a calculator available? Indeed, when you key a number into a calculator, how is the value of the sine, the square root, the logarithm or any other mathematical function determined?

Consider the particular case of calculating the value of $\sin x$ for a value of x (in radians) which lies in the interval $0 \leq x \leq \pi/2$.

We first meet the sine function as the sine of an angle, with $\sin x = a/c$ as in figure 2.1.

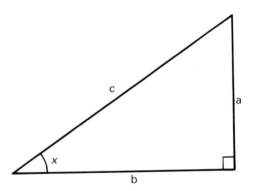

Figure 2.1

However this does not provide a practical or accurate means of calculating values of the function and another approach must be sought.

The graph of $y = \sin x$ for $0 \leq x \leq 2$ is shown in figure 2.2.

For small values of x, the curve looks as though it could be approximated by a line; in particular, by the line whose slope is the same as that of the graph of $y = \sin x$ at $x = 0$.

Since $y = \sin x$, $\dfrac{dy}{dx} = \cos x$ and when $x = 0$, $\dfrac{dy}{dx} = 1$.

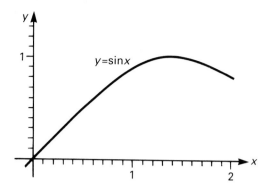

Figure 2.2

Hence the line $y = x$ will approximate the graph of $y = \sin x$ near $x = 0$. To investigate how good an approximation this gives, examine the graphs of $y = x$ and $y = \sin x$ displayed in figure 2.3 and the table of values of $\sin x$ and $|x - \sin x|$, each to 3 decimal places, which follows.

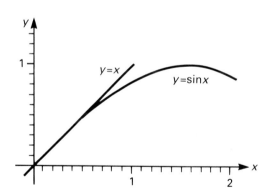

Figure 2.3

x	0.0	0.2	0.4	0.6	0.8	1.0
$\sin x$	0.000	0.199	0.389	0.565	0.717	0.841
$\lvert x - \sin x \rvert$	0.000	0.001	0.011	0.035	0.083	0.159

For small values of x, the approximation is clearly very good. However, as x increases, the approximation given by $y = x$ becomes progressively less satisfactory. Clearly $\sin x = x$ is a first approximation, but needs refinement if it is to be useful for other than very small values of x.

It might be possible to improve the approximation to $\sin x$ for $0 \le x \le 2$ by using a polynomial approximation of higher degree and matching higher order derivatives at $x = 0$. For example, using a quadratic polynomial

$$\sin x \approx a_0 + a_1 x + a_2 x^2$$

and matching the values of the polynomial and its first two derivatives at $x = 0$ with those of $\sin x$, ensures that the slope and the rate of change of the slope of the polynomial equal those of the sine function at $x = 0$. This gives the following results:

For the approximation to be exact at $x = 0$,

$$\sin x = a_0 + a_1 x + a_2 x^2 \qquad \text{when } x = 0 \quad \Rightarrow a_0 = 0;$$

for the first derivatives to be equal at $x = 0$

$$\cos x = a_1 + 2a_2 x \qquad \text{when } x = 0 \quad \Rightarrow a_1 = 1;$$

for the second derivatives to be equal at $x = 0$,

$$-\sin x = 2a_2 \qquad \text{when } x = 0 \quad \Rightarrow a_2 = 0.$$

Hence the quadratic approximation is

$$\sin x \approx 0 + x + 0x^2 = x$$

This has not improved on the linear approximation, but using

$$\sin x \approx a_0 + a_1 x + a_2 x^2 + a_3 x^3$$

gives, when $x = 0$,

$$\sin x = a_0 + a_1 x + a_2 + a_3 x^3 \qquad \Rightarrow a_0 = 0$$
$$\cos x = a_1 + 2a_2 x + 3a_3 x^2 \qquad \Rightarrow a_1 = 1$$
$$-\sin x = 2a_2 + 6a_3 x \qquad \Rightarrow a_2 = 0$$

$$-\cos x = 6a_3 \qquad \Rightarrow a_3 = -\frac{1}{6}$$

So the cubic approximation is

$$\sin x \approx x - \frac{1}{6} x^3.$$

The graphs of $y = \sin x$ and $y = x - \frac{1}{6} x^3$ are displayed in figure 2.4 and a table showing corresponding values of $\sin x$ and $x - \frac{1}{6} x^3$ follows.

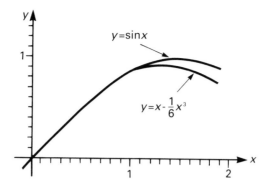

Figure 2.4

x	0.0	0.2	0.4	0.6	0.8	1.0
$\sin x$	0.000	0.199	0.389	0.565	0.717	0.841
$x - \dfrac{1}{6}x^3$	0.000	0.199	0.389	0.564	0.715	0.833
$\left\lvert x - \dfrac{1}{6}x^3 - \sin x \right\rvert$	0.000	0.000	0.000	0.001	0.002	0.008

In this case, the values given by the polynomial are the same as those of the function, correct to 3 decimal places, for $0 \leqslant x \leqslant 0.4$.

For greater accuracy, you could continue this process by looking at, say, the quintic approximation.

$$\sin x \approx a_0 + a_1 x + a_2 x^2 + a_3 x^3 + a_4 x^4 + a_5 x^5$$

where $a_0 = a_2 = 0$, $a_1 = 1$, $a_3 = -\dfrac{1}{6}$ and a_4 and a_5 are to be found.

Taylor Polynomials

The above example suggests a method for obtaining a general polynomial

$$a_0 + a_1 x + a_2 x^2 + \ldots + a_n x^n$$

NA

which could be used to approximate any function, f, for values of x near to $x = 0$. The only requirement is that the function f can be differentiated n times.

Given a suitable function f, the coefficients $a_0, a_1, a_2 \ldots a_n$ can be determined by equating the values of the polynomial and its derivatives with those of f and its derivatives as follows.

$$f(x) \approx a_0 + a_1 x + a_2 x^2 + \ldots + a_n x^n$$

so that

$$a_0 = f(0)$$

Taking first derivatives,

$$f'(x) \approx a_1 + 2a_2 x + 3a_3 x^2 + \ldots + n a_n x^{n-1}$$

so that

$$a_1 = f'(0)$$

Differentiating again,

$$f''(x) \approx 2a_2 + 3 \times 2a_3 x + 4 \times 3a_4 x^2 + \ldots + n(n-1) a_n x^{n-2}$$

so that $f''(0) \approx 2a_2$

giving $a_2 = \dfrac{1}{2} f''(0)$

Continuing in this way,

$$f''(x) \approx 3 \times 2a_3 + 4 \times 3 \times 2a_4 x + \ldots + n(n-1)(n-2) a_n x^{n-3}$$
$$f'''(0) = 3 \times 2a_3$$

giving $a_3 = \dfrac{1}{3 \times 2} f'''(0)$

and $a_4 = \dfrac{1}{4 \times 3 \times 2} f^{(iv)}(0) = \dfrac{1}{4!} f^{(iv)}(0)$

$$a_5 = \dfrac{1}{5!} f^{(v)}(0) \qquad \text{etc.}$$

so that $a_n = \dfrac{1}{n!} f^{(n)}(0).$

Therefore the polynomial which approximates $f(x)$ is

$$f(0) + x f'(0) + \frac{x^2}{2!} f''(0) + \ldots + \frac{x^n}{n!} f^{(n)}(0).$$

This polynomial is called *the nth degree Taylor polynomial of the function f at* $x = 0$.

EXAMPLE Obtain the Taylor polynomial of degree 3 which approximates the function

$$f(x) = \sqrt{1+x} \quad \text{near to } x = 0.$$

Solution:

The Taylor polynomial is

$$f(x) \approx f(0) + xf'(0) + \frac{x^2}{2!}f''(0) + \frac{x^3}{3!}f'''(0).$$

$$f(x) = (1+x)^{1/2} \qquad\qquad\qquad f(0) = 1$$

$$f'(x) = \frac{1}{2}(1+x)^{-1/2} \qquad\qquad\qquad f'(0) = \frac{1}{2}$$

$$f''(x) = \frac{1}{2}\left(-\frac{1}{2}\right)(1+x)^{-3/2} \qquad\qquad f''(0) = -\frac{1}{4}$$

$$f'''(x) = \left(-\frac{1}{4}\right)\left(-\frac{3}{2}\right)(1+x)^{-5/2} \qquad f'''(0) = \frac{3}{8}$$

Hence

$$f(x) \approx 1 + x \times \frac{1}{2} + \frac{x^2}{2} \times \left(-\frac{1}{4}\right) + \frac{x^3}{6} \times \frac{3}{8}$$

$$= 1 + \frac{1}{2}x - \frac{1}{8}x^2 + \frac{1}{16}x^3.$$

NOTES

1. If a higher degree Taylor polynomial were required, we would simply calculate $f^{(iv)}(0)$ and then

$$f(x) \approx 1 + \frac{1}{2}x - \frac{1}{8}x^2 + \frac{1}{16}x^3 + \frac{1}{4!}f^{(iv)}(0)x^4.$$

2. The binomial expansion could be used to give a series expansion for $\sqrt{1+x}$ for values of x in the interval $-1 < x < 1$. You should verify that the Taylor polynomial of degree 3 consists of the first 4 terms of the binomial expansion of $\sqrt{1+x}$.

Q

1. Find the Taylor polynomial of degree 4 generated by the function $f(x) = \cos x$ at $x = 0$.

2. Determine the Taylor polynomials of degree 4 generated by each of the functions

   (i) $\dfrac{1}{1+x}$ and (ii) $\dfrac{1}{(1+x)^2}$ at $x = 0$.

In each case, use your calculator to calculate the error which arises from using the polynomial as an approximation to the function at $x = 0.1, 0.5$ and 1.0. For each function, use a graph plotter to display the graph of the function and the corresponding polynomial for $-0.5 < x < 2$.

3. Find the Taylor polynomial of degree 3 generated by the function $f(x) = \arc \sin x$ at $x = 0$. Denote this polynomial by $P_3(x)$. Calculate the error made in approximating $f(x)$ by $P_3(x)$ at $x = 0.1, 0.4$ and 0.7. Does the value of $P_3(x)$ for $x > 1$ have any significance?

In the above examples, we obtained Taylor polynomials which approximated a given function near to $x = 0$. As x moves away from the value $x = 0$, the approximations become less accurate. A more general result gives Taylor polynomials at any point $x = a$ at which the function can be differentiated.

To illustrate how this result can be obtained, consider how we could obtain a polynomial which approximates $\sin x$ near to $x = 1.5$ (figure 2.5).

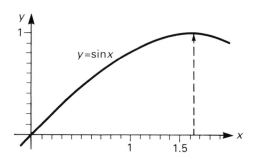

Figure 2.5

Since the values of interest lie near to $x = 1.5$ it might be better to consider values of

$$y = \sin(1.5 + h)$$

where h is small. That is, to consider y as a function of h (figure 2.6).

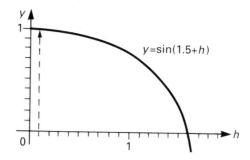

Figure 2.6

It is now appropriate to use Taylor polynomials to give an approximation to $\sin(1.5 + h)$.

Since $\dfrac{d}{dh} \sin(1.5 + h) \quad = \quad \cos(1.5 + h) \quad = \quad \cos 1.5$ when $h = 0$,

$\dfrac{d^2}{dh^2} \sin(1.5 + h) \quad = -\sin(1.5 + h) \quad = -\sin 1.5$ when $h = 0$

and $\dfrac{d^3}{dh^3} \sin(1.5 + h) \quad = -\cos(1.5 + h) \quad = -\cos 1.5$ when $h = 0$

the third degree Taylor approximation is

$$\sin(1.5 + h) \approx \sin 1.5 + h \cos 1.5 + \frac{h^2}{2!}(-\sin 1.5) + \frac{h^3}{3!}(-\cos 1.5).$$

In general if we wish to expand a function, f, of x in an interval about the point $x = a$,

letting $\qquad x = a + h$, where a is a constant,

gives $\qquad \dfrac{dx}{dh} = 1$

so that $\qquad \dfrac{d}{dh} f(a + h) = \dfrac{d}{dx} f(x) \times \dfrac{dx}{dh} = f'(x)$

Hence, when $h = 0$, $\dfrac{d}{dh} f(a) = f'(a)$; this gives

$$f(a + h) \approx f(a) + hf'(a) + \frac{h^2}{2!}f''(a) + \frac{h^3}{3!}f'''(a).$$

This is the Taylor polynomial of degree 3 which approximates $f(x)$ near to $x = a$.

EXAMPLE

Use the Taylor polynomial of degree 3 about the point $x = \pi$ to obtain an approximation to the value of $\sin 3$.

Solution:
The Taylor polynomial is

$$f(a + h) \approx f(a) + hf'(a) + \frac{h^2}{2!}f''(a) + \frac{h^3}{3!}f'''(a)$$

In this example, $a = \pi$

and

$$
\begin{aligned}
f(x) &= \sin x & f(\pi) &= 0 \\
f'(x) &= \cos x & f'(\pi) &= -1 \\
f''(x) &= -\sin x & f''(\pi) &= 0 \\
f'''(x) &= -\cos x & f'''(\pi) &= 1
\end{aligned}
$$

Hence

$$f(\pi + h) \approx 0 - h + 0 + \frac{1}{6}h^3.$$

To approximate $\sin 3$,

$$\pi + h = 3 \quad \text{so} \quad h = 3 - \pi$$
$$= -0.1416$$

Hence

$$f(3) \approx -(-0.1416) + \frac{1}{6}(-0.1416)^3$$

$$= 0.1411.$$

NOTE

The value of $\sin 3$ given by a calculator is 0.1411 correct to 4 decimal places.

The general form of the Taylor polynomial of degree n for values of x near to $x = 0$ is

$$f(a + h) \approx f(a) + hf'(a) + \frac{h^2}{2!}f''(a) + \ldots + \frac{h^n}{n!}f^{(n)}(a).$$

Instead of using the value of h in this expression we could set $x = a + h$ so that $h = x - a$ to give the alternative form of the polynomial approximation.

$$f(x) \approx f(a) + (x - a)f'(a) + \frac{(x - a)^2}{2!}f''(a) + \ldots + \frac{(x - a)^n}{n!}f^{(n)}(a).$$

Both forms give useful expressions with which to approximate the function $f(x)$ near to $x = a$.

Q

1. Obtain the Taylor polynomial of degree 3, $P_3(x)$, which approximates the function $\sin x$ near to $x = \pi$. Use a graph plotter to show the graph of $y = \sin x$ superimposed on the graph of $y = P_3(x)$. Estimate, from the graphs, the values of x for which $|\sin x - P_3(x)| < 0.1$.

2. Determine the Taylor polynomial of degree 3 for the function

$$f(x) = \frac{1}{x} \text{ at } x = 1.$$

Does this function have Taylor polynomials at $x = 0$? Explain your answer.

Taylor Series

If a function f has derivatives of all orders at a point $x = a$, then there is no bound to the degree of Taylor polynomial which can be formed. From the examples considered in earlier sections, it appears that the higher the degree of the Taylor polynomial, the better the approximation it gives to the function f near to $x = a$. The infinite series

$$f(a) + (x - a)f'(a) + \frac{(x - a)^2}{2!}f''(a) + \ldots$$

is called the Taylor series generated by the function f at the point $x = a$. From the way in which the polynomials were derived, we know that the sum of n terms of this series, $s_n(x)$, is the Taylor polynomial $p_n(x)$ which has the property that $p_n(x)$ and its first n derivatives equal the corresponding values of $f(x)$ and its derivatives at $x = a$. However we have not *proved* that for other values of x, the values of $s_n(x)$ will approximate $f(x)$. The general result is named after the English mathematician Brook Taylor; it is known as Taylor's Theorem and states that if the function f and its derivatives can be found for all values of x lying between a and b and $p_n(x)$ is the nth degree Taylor polynomial at $x = a$ then

$$f(b) = p_n(b) + \text{error}$$

where the error term can be specified (see figure 2.7).

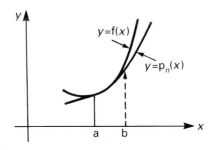

Figure 2.7

The special case of Taylor's series in which $a = 0$, namely, the series

$$f(0) + xf'(0) + \frac{x^2}{2!}f''(0) + \frac{x^3}{3!}f'''(0) + \ldots$$

is known as Maclaurin's series.

HISTORICAL NOTE

Brook Taylor, a great admirer of Newton, worked at Cambridge in the early years of the 18th century. He is remembered for his work on Taylor polynomials although the significance of the work was not appreciated for another hundred years when the error terms were formulated. In fact, in 1688 the Scottish mathematician James Gregory had, unknown to Taylor, already published the series expansions for functions. Colin Maclaurin, another Scot, was a prolific mathematician who went to the University of Glasgow at the age of 11 and became Professor of Mathematics at Aberdeen at the age of 19. He supported the crown in the Jacobite uprising and had to flee south, dying in exile in York in 1746. Although he made no claim to having discovered the results, Taylor expansions about $x = 0$ are known as Maclaurin's series.

EXAMPLE

Obtain the first four terms in the Taylor series expansion of the function $f(x) = e^x$ about the point $x = 1$. Hence obtain the first, second and third degree approximation to the value of $e^{1.1}$.

Solution:
The first four terms of the Taylor expansion are

$$f(1) + (x - 1)f'(1) + \frac{(x - 1)^2}{2!}f''(1) + \frac{(x - 1)^3}{3!}f'''(1)$$

Now $f(x) = f'(x) = f''(x) = f'''(x) = e^x$ and $e^1 = 2.71828.$

Hence the required Taylor expansion is

$$2.71828\left(1 + (x - 1) + \frac{(x - 1)^2}{2!} + \frac{(x - 1)^3}{3!}\right)$$

It follows that the approximations to $e^{1.1}$ are:

First degree:	$2.71828 + 0.271818 = 2.990108$
Second degree:	$2.990108 + 0.013591 = 3.003699$
Third degree:	$3.003699 + 0.000453 = 3.004152$

Using a calculator, we find that the value of $e^{1.1}$, correct to six decimal places, is 3.004166, showing that the approximation given by the third degree approximation is correct to 4 decimal places. Indeed, noting the errors in each approximation gives the following table.

Degree of approximation n	$\vert f(1.1) - P_n(1.1) \vert$
1	0.014058
2	0.000467
3	0.000014

Successive terms in the Taylor series are decreasing; looking at the error in the first degree approximation, it is seen to be close to the value of the third term in the Taylor series, namely 0.013591.

Similarly, the error in the second degree approximation is close to the value of the fourth term, namely 0.000453.

Indeed, Taylor's Theorem states that if the first degree approximation is used,

$$f(x) = f(a) + (x - a)f'(a) + \text{error}$$

where $\quad\quad \text{error} = \dfrac{(x - a)^2}{2!} f''(\eta) \quad$ where $a < \eta < x.$

The error $\dfrac{(x - a)^2}{2!} f''(\eta)$ introduced by approximating $f(x)$ by the first degree Taylor polynomial $P_1(x)$ is called a *truncation error*; it is caused by *truncating* or cutting short the expansion at the term in $(x - a)$.

In general, for an nth degree Taylor approximation, the error is given by

$$\dfrac{(x - a)^{n+1}}{(n + 1)!} f^{(n+1)}(\eta) \quad \text{where } a < \eta < x.$$

Although it is not possible to determine the exact value of the error term since the value of η is not known, the result can often be used to put an upper limit on the size of the error in the approximation.

For example, the first order Taylor expansion of $\sin x$ about $x = 0$ is

$$p_1(x) = x \quad \text{with the error given by } \dfrac{x^2}{2!} f''(\eta) \text{ where } 0 < \eta < x.$$

Since $f(x) = \sin x$, $f'(x) = \cos x$ and $f''(x) = -\sin x$, the error is

$$\dfrac{x^2}{2}(-\sin \eta) \text{ where } 0 < \eta < x.$$

When $x = 0.2$, the magnitude of the error is

$$\left\vert \dfrac{x^2}{2} \sin \eta \right\vert = \left\vert \dfrac{(0.2)^2}{2} \sin \eta \right\vert, \text{ with } 0 < \eta < 0.2$$

$$< \dfrac{0.04}{2} \times 0.2 = 0.004, \text{ since } \vert \sin 0.2 \vert < 0.2.$$

When $x = 0.4$, the magnitude of the error $< \dfrac{(0.4)^2}{2} \times 0.4 = 0.032$.

Clearly as x increases, the magnitude of the error term increases.

$\boxed{Q}$ The first degree Taylor expansion of $f(x) = \sqrt{1+x}$ about $x = 0$ is $1 + \dfrac{x}{2}$.

Use the formula given above to estimate the truncation error in using this polynomial as an approximation to $f(x)$ at (i) $x = 0.01$, and (ii) $x = 0.1$.

At the start of the chapter, the problem of how a calculator obtains values of mathematical functions was posed. Taylor's theorem tells us that, as long as certain conditions on the function and its derivatives are met, the function can be approximated by a polynomial function. This is clearly a valuable result.

Further work on function approximation has produced other series expansions which can approximate functions more efficiently and it is these series which are used in modern computing.

It should always be remembered that functions can only be approximated by Taylor polynomials if the derivatives of the function exist for an interval which contains the values of x in which we are interested. For example, it would not be possible to obtain a polynomial approximation to $f(x) = \tan x$ for values of x near to $x = \pi/2$ (figure 2.8) nor for the function shown in figure 2.9 near to $x = 1$.

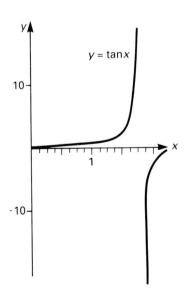

Figure 2.8

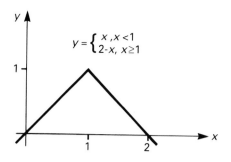

$$y = \begin{cases} x, x < 1 \\ 2-x, \ x \geq 1 \end{cases}$$

Figure 2.9

Curve Fitting

Taylor polynomials provide a useful method for approximating a given function, f, by a polynomial. However, in some cases where the function is not defined explicitly, we would still like to approximate the function using a polynomial.

For example, in a project to design an aerofoil, a designer produces the shape shown in figure 2.10. If the design and manufacture are to be computerised, information on the shape must be passed to other parts of the computer system.

Figure 2.10

It may also be the case that the area of the shape is required. Each of these requirements could be met if it were possible to approximate the curved part of the perimeter with a polynomial function.

In this case, the function f is not known explicitly but we could obtain as many points $(x_r, f(x_r))$ as we wish, which lie on the curve in figure 2.11. Problems of this type are called *curve fitting* problems.

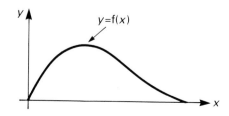

Figure 2.11

A similar problem arises if we carry out an experiment in which a set of readings (x_r, y_r), $r = 0, 1, 2, \ldots, n$, is taken and we wish to approximate the smooth curve which joins the points. The approximating polynomial might be used to give an estimate of the value of y corresponding to a non-tabular point x as in figure 2.12.

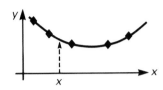

Figure 2.12

If x lies between the smallest and largest values of x_r, this process is called *interpolation* and, for this reason, the polynomial approximation to the function is called an *interpolating polynomial*.

EXAMPLE Determine the second degree polynomial which passes through $(1, 1)$, $(10, 0.1)$ and $(20, 0.05)$.

Solution:
Let the polynomial be p where

$$p(x) = a_0 + a_1 x + a_2 x^2$$

We must determine the coefficients a_0, a_1 and a_2.

Since $(1, 1)$ lies on $y = p(x)$,

$$1 = a_0 + a_1 + a_2 \qquad (1)$$

Similarly, $(10, 0.1)$ and $(20, 0.05)$ lie on the curve so

$$0.1 = a_0 + 10a_1 + 100a_2 \qquad (2)$$

and $\qquad\qquad 0.05 = a_0 + 20a_1 + 400a_2 \qquad (3)$

These three equations in a_0, a_1 and a_2 can be solved by taking

Equation (2) − Equation (1), $\qquad -0.9 = 9a_1 + 99a_2$
Equation (3) − Equation (1), $\qquad -0.95 = 19a_1 + 399a_2$

and solving these equations gives

$$a_0 = 1.15, \quad a_1 = -0.155, \quad a_2 = 0.005$$

Hence the interpolating polynomial is

$$p(x) = 1.15 - 0.155x + 0.005x^2$$

The graph of $y = p(x)$ is shown in figure 2.13.

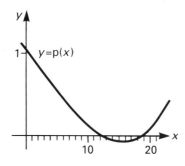

Figure 2.13

Note that only one quadratic curve passes through the three given points. Extending this, only one line passes through two points, only one cubic curve passes through a given set of four points and, in general, only one curve of degree n,

$$y = a_0 + a_1 x + a_2 x^2 + \ldots + a_n x^n$$

will pass through a given set of $n + 1$ points.

It is always possible to determine the interpolating polynomial of degree n by solving a set of $n + 1$ equations in the unknowns $a_0, a_1, a_2 \ldots a_n$. If n is large, this can be laborious and if the polynomial is to be used to interpolate a particular function value, a better method of obtaining the function value is due to Lagrange.

Lagrange Interpolation Polynomials

HISTORICAL NOTE

Joseph-Louis Lagrange (1736–1813) had an Italian mother and French father and is reputed to have been appointed professor of mathematics at Turin when he was aged 16. He was the only one of his parents' eleven children to survive infancy and by middle age, he suffered from a depressive illness. His main work was in the field of analysis where one problem he worked on was that of finding stationary points of functions of several variables. He also did work on mechanics where he advised Napoleon on the setting up of the metric system of measurement. He was a specialist who used analytic rather than geometric methods to solve problems in mechanics and, at the start of one of his treatises he claimed 'No diagrams will be found in this work'. Lagrange was honoured by the King of Sardinia, Frederick the Great and, most lavishly, by Napoleon.

Consider the determination of the cubic polynomial, $p_3(x)$, which passes through the four points $A(x_0, y_0)$, $B(x_1, y_1)$, $C(x_2, y_2)$ and $D(x_3, y_3)$ as shown in figure 2.14.

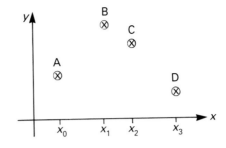

Figure 2.14

As a first step, consider the function $l_0(x)$ defined by

$$l_0(x) = \frac{(x - x_1)(x - x_2)(x - x_3)}{(x_0 - x_1)(x_0 - x_2)(x_0 - x_3)}.$$

Notice that $l_0(x)$ is a cubic polynomial in x and, when $x = x_0$, we see that $l_0(x_0) = 1$; also, $l_0(x_1) = l_0(x_2) = l_0(x_3) = 0$. If follows that the curve $y = l_0(x)y_0$ passes through (x_0, y_0) and the points $(x_1, 0)$, $(x_2, 0)$ and $(x_3, 0)$.

Similarly, if $l_1(x) = \dfrac{(x - x_0)(x - x_2)(x - x_3)}{(x_1 - x_0)(x_1 - x_2)(x_1 - x_3)}$,

$l_1(x)$ is a cubic polynomial in x with $l_1(x_1) = 1$ and $l_1(x_0) = l_1(x_2) = l_1(x_3) = 0$. Also the curve $y = l_1(x)y_1$, passes through (x_1, y_1) and $(x_0, 0)$, $(x_2, 0)$ and $(x_3, 0)$. Similar functions $l_2(x)$ and $l_3(x)$ can be defined with $l_2(x_2) = l_3(x_3) = 1$ and $l_2(x_0) = l_2(x_1) = l_2(x_3) = 0$ so that $l_3(x_0) = l_3(x_1) = l_3(x_2) = 0$.

Consider the cubic function p_3 defined by

$$p_3(x) = l_0(x)y_0 + l_1(x)y_1 + l_2(x)y_2 + l_3(x)y_3$$

We note that $\quad p_3(x_0) = y_0, \ p_3(x_1) = y_1, \ p_3(x_2) = y_2.$

The curve $y = p_3(x)$ passes through the four points A, B, C and D, and so it must be the unique interpolating polynomial of degree 3.

To illustrate the use of this formulation of the cubic polynomial when it is required for interpolation, suppose that we are given the four points $(1, 12)$, $(2, 15)$, $(5, 25)$ and $(6, 30)$ and we want to approximate the function value when $x = 4$.

We know that

$$p_3(x) = l_0(x)y_0 + l_1(x)y_1 + l_2(x)y_2 + l_3(x)y_3$$

where
$$l_0(x) = \frac{(x-x_1)(x-x_2)(x-x_3)}{(x_0-x_1)(x_0-x_2)(x_0-x_3)} = \frac{(x-2)(x-5)(x-6)}{(1-2)(1-5)(1-6)}$$

$$= -\frac{1}{20}(x-2)(x-5)(x-6)$$

$$l_1(x) = \frac{(x-x_0)(x-x_2)(x-x_3)}{(x_1-x_0)(x_1-x_2)(x_1-x_3)} = \frac{(x-1)(x-5)(x-6)}{(2-1)(2-5)(2-6)}$$

$$= \frac{1}{12}(x-1)(x-5)(x-6)$$

$$l_2(x) = \frac{(x-x_0)(x-x_1)(x-x_3)}{(x_2-x_0)(x_2-x_1)(x_2-x_3)} = \frac{(x-1)(x-2)(x-6)}{(5-1)(5-2)(5-6)}$$

$$= -\frac{1}{12}(x-1)(x-2)(x-6)$$

$$l_3(x) = \frac{(x-x_0)(x-x_1)(x-x_2)}{(x_3-x_0)(x_3-x_1)(x_3-x_2)} = \frac{(x-1)(x-2)(x-5)}{(6-1)(6-2)(6-5)}$$

$$= \frac{1}{20}(x-1)(x-2)(x-5)$$

We require the value of $p_3(4) = l_0(4)y_0 + l_1(4)y_1 + l_2(4)y_2 + l_3(4)y_3$

Now,
$$l_0(4) = -\frac{1}{20}(4) = -\frac{1}{5}$$

$$l_1(4) = \frac{1}{12}(6) = \frac{1}{2}, \quad l_2(4) = -\frac{1}{12}(-12) = 1$$

$$l_3(4) = \frac{1}{20}(-6) = -\frac{3}{10}$$

so that
$$p_3(4) = -\frac{1}{5}y_0 + \frac{1}{2}y_1 + y_2 - \frac{3}{10}y_3$$

$$= -\frac{1}{5} \times 12 + \frac{1}{2} \times 15 + 25 - \frac{3}{10} \times 30$$

$$= 21.1.$$

Note that the values of y_0, y_1, y_2 and y_3 are not used until the final step and the values of $l_r(4)$, for $r = 0, 1, 2,$ and 3, namely

$$l_0(4) = -\frac{1}{5}, \quad l_1(4) = \frac{1}{2}, \quad l_2(4) = 1, \quad l_3(4) = -\frac{3}{10}$$

can be considered as *weighting factors* which determine the *weight* to be given to each of the y values. Note that the data point nearest to $x = 4$ is the point (5, 25) and the weighting factor associated with this point has the largest magnitude.

Q The values of a function f are given in the following table.

x	1	2	4
$f(x)$	1.00	0.0	1.5

Obtain the Lagrange interpolating polynomial of degree 2 for the given data. Use this polynomial to estimate f(2.5).

In general, the nth degree Lagrange interpolating polynomial is

$$p_n(x) = \sum_{k=0}^{n} l_k(x) y_k$$

where

$$l_k(x) = \frac{(x - x_0)(x - x_1) \ldots (x - x_{k-1})(x - x_{k+1}) \ldots (x - x_n)}{(x_k - x_0)(x_k - x_1) \ldots (x_k - x_{k-1})(x_k - x_{k+1}) \ldots (x_k - x_n)}$$

$$= \prod_{\substack{i=0 \\ i \neq k}}^{n} \frac{x - x_i}{x_k - x_i}.$$

The symbol Π is used to represent the *product* of the factors of the form

$$\frac{x - x_i}{x_k - x_i} \text{ with } i = 0, 1, 2, \ldots, k-1, k+1, \ldots n.$$

Some points to note when using interpolating polynomials are illustrated in the following example.

EXAMPLE In an earlier example we obtained the quadratic polynomial

$$p_2(x) = 1.15 - 0.155x + 0.005x^2$$

which passes through the points (1, 1), (10, 0.1) and (20, 0.05). Use $p_2(x)$ and also the polynomial $p_1(x)$ through (10, 0.1) and (20, 0.05) to give estimates of the value of the function at $x = 16$.

Given that the data was obtained from the function f with $f(x) = \frac{1}{x}$, comment on the approximations you have obtained for f(16).

Solution:

$$p_2(16) = 1.15 - 0.155 \times 16 + 0.005 \times 16^2$$
$$\approx -0.05$$

Also
$$p_1(x) = \frac{x - x_1}{x_0 - x_1} y_0 + \frac{x - x_0}{x_1 - x_0} y_1$$

$$= \frac{x - 20}{10 - 20} \times 0.1 + \frac{x - 10}{20 - 10} \times 0.05$$

$$= 0.15 - 0.005x$$

giving
$$p_1(16) = 0.07.$$

Now
$$f(16) = \frac{1}{16} = 0.0625$$

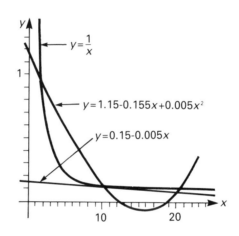

Figure 2.15

From the diagram above, not only is $p_2(16) < 0$ but the slope of the curve $y = p_2(x)$ does not approximate the slope of the curve $y = f(x)$ for $10 < x < 20$. In examples in which a large number of function values are available, it is not always best to use all of them. Not only is the calculation lengthy but if a polynomial of degree n is used for interpolation, the derivative of the polynomial will have degree $n - 1$ so that there could be as many as $n - 1$ turning-points on $y = o_n(x)$, some of which may lie in the interval in which we are interested.

This problem is illustrated in figure 2.16 in which the interpolating polynomial has two turning-points at P and Q which we would not have expected in the function being approximated.

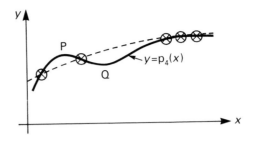

Figure 2.16

Another point to note from the above example is that although the line

$$y = 0.15 - 0.005x$$

approximates the function f well for $10 < x < 20$ it gives a very poor approximation for values of x less than 10. Since the points used to obtain $p_1(x)$ were (10, 0.1) and (20, 0.05), when $p_1(x)$ is used to approximate f for $x < 10$, we are not interpolating but *extrapolating*. In extrapolation, the information about f being used to give the approximating polynomial pertains to an interval on the x-axis which does *not* contain the value of x at which the approximation is required; this is frequently an unreliable procedure to use.

Investigation

Investigate the use of Lagrange polynomials of different degrees to interpolate the following data.

1. Values of $(x_i, f(x_i))$ are given in the table below.

i	0	1	2	3	4
x_i	0	1	8	27	64
$f(x_i)$	0	1	2	3	4

Use the points (27, 3) and (8, 2) to estimate the value of f(20).

Use Lagrange polynomials of degrees 2, 3 and 4 to estimate f(20). Explain your choice of data points used to determine each polynomial.

2. Given that $f(x) = x^{1/3}$, calculate the % error in each estimate of f(20).

Investigation continued

3. Express the linear and quadratic polynomials in the form $a_0 + a_1 x$ and $a_0 + a_1 x + a_2 x^2$ respectively. The cubic polynomial obtained using f(0), f(1), f(8) and f(27) is

$$p_3(x) = 0.00384x^3 - 0.142x^2 + 1.138x$$

(You could verify this result.)
Use a graph plotting program to sketch each of these polynomial functions.
Copy each graph into your report, showing the data points used.

Before concluding this section, we should add a cautionary note about the interpolating polynomial. Suppose that, for a function f, we are given a set of function values $(x_r, f(x_r))$, $r = 0, 1, 2, \ldots$ which can be used to obtain polynomials of different degrees which interpolate a *subset* of points.

For example, given the four points A, B, C and D, shown in figure 2.17, polynomial functions $y = g(x)$ and $y = h(x)$, of different degrees, can be obtained which approximate the values of f(x), for $x_0 \leqslant x \leqslant x_3$.

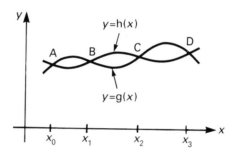

Figure 2.17

However, looking at the point A, on $y = g(x)$, where the slope is positive, the value of $g'(x_0)$ is positive and is not approximately equal to the value of $h'(x_0)$ which is negative. In general, when approximating a function, f, we can be much less confident about the approximation this gives for the derived function f'.

Exercise 2

1. Obtain the Taylor polynomials of degree 4 for the following functions f at $x = 0$

   (i) $f(x) = \ln(1 + x)$ (ii) $f(x) = e^{2x}$ (iii) $f(x) = \tan x$
   (iv) $f(x) = \sin x + \cos x$.

2. Obtain the Taylor polynomials of degree 3 for the following functions f at the given points

   (i) $f(x) = x^2 \ln x$ at $x = 5$
   (ii) $f(x) = \cos x$ at $x = \pi/3$

Exercise 2 continued

    (iii) $f(x) = \sqrt[3]{x}$ at $x = 8$

    (iv) $f(x) = \ln(\cos x)$ at $x = \pi/3$.

3. Using a graph plotter, sketch the graph of the sine function for $-1 \leqslant x \leqslant 6$. Obtain Taylor polynomials of degree 4 which approximate the sine function

    (i) near to $x = 0$ (ii) near to $x = \pi/2$ (iii) near to $x = \pi$

    Superimpose the graphs of each of these polynomials in turn on the original graph, checking that they approximate the given function in the appropriate region.

4. We know that $\dfrac{d}{dx} \sin x = \cos x$ and $\displaystyle\int \sin x \, dx = -\cos x + C$. Obtain the Taylor series expansions for the sin and cosine functions and investigate whether or not the above results hold for the series approximations.

5. Find the Taylor polynomial of degree 3 for the function f expanded about $x = 1$ where $f(x) = e^{-x}$. Use this polynomial to approximate $f(0.99)$. Use your calculator to evaluate $f(0.99)$ and write down the number of correct decimal places in the approximation.

6. Each of the following functions is approximated using the first degree Taylor polynomial at $x = 0$.

    (i) $\sin x$, approximated at $x = \pi/4$

    (ii) e^x at $x = 0.1$.

    Use the formula for the error term to determine a maximum bound for the error in the approximation. Compare this maximum error with the actual error, using your calculator to give the function values.

7. The functions $\sin x$ and $\tan x$ are approximated near $x = 0$, using the first degree Taylor polynomials. By examining the error term in each approximation determine which approximation you would expect to fit the given function better. Check your answer by showing each function and its Taylor polynomial on a graph.

8. Let $x = x_0$ be a first approximation to a root of the equation $f(x) = 0$. Using the first two terms of the Taylor expansion of $f(x_0 + h)$, determine an expression for h such that $x_1 = x_0 + h$ is an improved solution of the equation. Verify that this gives an alternative derivation of the Newton-Raphson method for solving $f(x) = 0$.

9. Given the following table of values of a function f, interpolate $f(2.3)$ using Lagrange interpolation.

x	1.1	1.7	3.0
$f(x)$	10.6	15.2	20.3

10. Use the method of Lagrange to obtain the parabola which passes through the following 3 points

x	0.5	1.2	3.1
y	-3.2	1.6	-1.8

Use this polynomial to estimate the maximum value of the function in the interval $(0.5, 3.1)$.

11. Determine the cubic polynomial which passes through the four points

$$(-2, 15), \ (-1, 0), \ (1, 0) \ \text{and} \ (2, 27)$$

and hence estimate the points on the curve at which $x = 0$ and $x = 3$. Comment on the reliability of the two estimates.

12. **Write down the values of** $f(x) = 1/x$ for $x = 2, 2.5$ and 4. Use the Lagrange interpolation formula to obtain a quadratic approximation $p_2(x)$, which approximates $f(x)$ for $2 \leqslant x \leqslant 4$. Compare the values of $f(x)$ and $p_2(x)$ for $x = 1, 3$ and 5. Compare also the values of $f'(x)$ and $p_2'(x)$ for the same values of x. Comment on your findings.

Does $\displaystyle\int_{2}^{4} p_2(x)\, dx$ give a good approximation to $\displaystyle\int_{2}^{4} f(x)\, dx$?

13. Use linear interpolation on the following two values of the logarithmic function to compute a table showing approximations to $\ln x$ for $x = 1, 2, 3 \ldots 10$.

Data values: $\ln 2 = 0.69315$
$\ln 5 = 1.60944$

Use your calculator to give the corresponding values of $\ln x$ and add two extra columns to the table to show the magnitude of the error, $|\varepsilon|$, and the magnitude of the relative error, $\dfrac{|\varepsilon|}{|\text{exact value}|}$,

in each interpolated or extrapolated value.

Comment on the values obtained in the table.

14. The value of the definite integral $\displaystyle\int_{1}^{2} e^{-x^2}\, dx$ is to be estimated. By

approximating e^{-x^2} by a cubic polynomial in the interval $(1, 2)$, obtain an estimate of the value of the definite integral.

Extended Question

1. Choose four values of x in the interval $(0, \pi)$ and obtain the corresponding values of $\sin x$, correct to 3 decimal places. Obtain the Lagrange interpolating polynomial of degree 3 which fits these data. Investigate how closely the area under the cubic polynomial between $x = 0$ and $x = \pi$ approximates the corresponding area under the sine curve.

The function $f(x) = \sin x$ in the interval $(0, \pi)$ could be approximated by a Taylor expansion about $x = \pi/2$. Obtain the fifth degree Taylor approximation to f about $x = \pi/2$. Use the formula for the error in the Taylor approximation to estimate the error in using this approximation to $\sin \pi$.

How well does the area under this polynomial approximation to $f(x)$ in $0 \leq x \leq \pi$ approximate the corresponding area under the sine curve?

KEY POINTS

- A known function f can be approximated in an interval about $x = 0$ by a Taylor polynomial $p_n(x)$, where

$$p_n(x) = f(0) + xf'(0) + \frac{x^2}{2!} f''(0) + \ldots + \frac{x^n}{n!} f^{(n)}(0).$$

- More generally, in an interval around $x = a$,

$$f(x) \approx f(a) + (x - a)f'(a) + \frac{(x - a)^2}{2!} f''(a) + \ldots + \frac{(x - a)^n}{n!} f^{(n)}(a).$$

- For a first degree Taylor approximation to a function f,

$$f(x) = f(a) + (x - a)f'(a) + \text{truncation error}$$

where the **truncation error** is given by

$$\frac{(x - a)^2}{2!} f''(\eta) \text{ with } a < \eta < x.$$

- If a function f is only specified by a set of points (x_0, y_0), (x_1, y_1) $\ldots (x_n, y_n)$, with $x_0 < x_1 < \ldots < x_n$, then the Lagrange Interpolation polynomial can be used to approximate the function as follows

$$p_n(x) = \sum_{k=0}^{n} l_k(x) y_k \quad \text{where} \quad l_k(x) = \prod_{\substack{i=0 \\ i \neq k}}^{n} \frac{x - x_i}{x_k - x_i}.$$

3 Numerical Integration

In the design of our buildings and of manufactured goods, irregular shapes are often used for aesthetic or practical reasons. Since properties of these shapes such as the length of a curve, an area or a volume may be required, we must look for mathematical techniques which could be used to calculate them. To illustrate such a technique, consider the following problem about the construction of a patio.

In a south-facing garden, it is decided that a patio should be constructed in a sheltered corner by the house. An architect produced the design shown in figure 3.1.

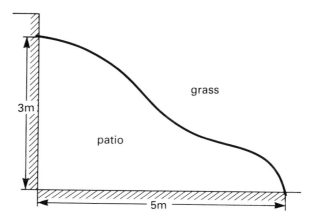

Figure 3.1

The patio is to be covered in concrete to a depth of 80 mm, the concrete being supplied already mixed. The problem is to estimate the volume of concrete which should be ordered. Since the volume is given by the product of the area of the patio and the depth of the concrete, the problem reduces to one of determining the area of the patio.

We realise that, if the function, f, which describes the curved edge of the patio, as shown in figure 3.2 were known, the area would be given by $\int_0^5 f(x)\,dx$.

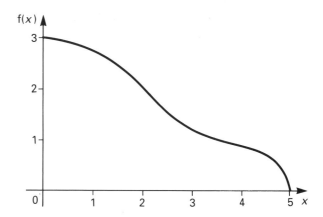

Figure 3.2

So, one approach would be to use the techniques of Chapter 2 to obtain a polynomial function, $p_n(x)$, which approximates the curve and then to approximate the area with $\int_0^5 p_n(x)\,dx$. However, a more direct method would be to subdivide the area into regular shapes, for example rectangles or triangles, which approximately cover the area. One of the ways in which this might be done underlies the following method.

The Trapezium Rule

Consider the computation of the definite integral which gives the area A, where $A = \int_a^b f(x)\,dx$.

The area could be divided into a number of strips of equal width by drawing lines parallel to the y-axis, as shown in figure 3.3.

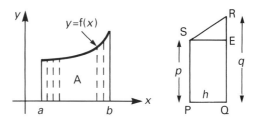

Figure 3.3

If the strips are narrow, each strip can be approximated by a *trapezium*, ie. a quadrilateral with one pair of parallel sides. If the lengths of the parallel sides are p and q and the width of the strip is h

$$\text{area of PQRS} = \frac{1}{2}h(p+q)$$

We can use this result to estimate the area A, if each strip has width h and the rth trapezium, T_r, has heights f_{r+1} and f_r as shown in figure 3.4.

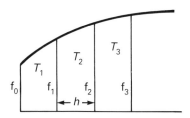

Figure 3.4

So
$$T_1 + T_2 + \ldots + T_n = \frac{1}{2}h(f_0 + f_1) + \frac{1}{2}h(f_1 + f_2) + \ldots + \frac{1}{2}h(f_{n-1} + f_n)$$

$$= \frac{1}{2}h(f_0 + 2f_1 + 2f_2 + \ldots + 2f_{n-1} + f_n)$$

$$= \frac{h}{2}(f_0 + f_n + 2(f_1 + f_2 + \ldots + f_{n-1}))$$

The result

$$\int_a^b f(x)\,dx \approx \frac{h}{2}(f_0 + f_n + 2(f_1 + f_2 + \ldots + f_{n-1}))$$

is known as the *trapezium rule*.

EXAMPLE Use the following data to estimate $\int_1^5 f(x)\,dx$ using the trapezium rule.

x	1.0	1.5	2.0	2.5	3.0	3.5	4.0	4.5	5.0
$f(x)$	0.00	0.41	0.69	0.92	1.10	1.25	1.39	1.50	1.61

Solution:

It is good practice to set down the calculation in a form which can be easily checked. One possible layout is shown in the following table.

x	$f(x)$	Factor	Product
1.0	0.00	1	0.00
1.5	0.41	2	0.82
2.0	0.69	2	1.38
2.5	0.92	2	1.84
3.0	1.10	2	2.20
3.5	1.25	2	2.50
4.0	1.39	2	2.78
4.5	1.50	2	3.00
5.0	1.61	1	1.61
			16.13

$$h = 0.5 \text{ and so } \int_1^5 f(x)\,dx \approx \frac{0.5}{2} \times 16.13 = 4.03$$

Q

1. Two points, A and B, on the x-axis, 2 units apart are shown below (all strips are of equal width). The values of f_i for $i = 0, 1, 2, 3, 4$ are given in the following table.

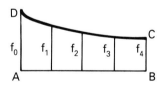

i	0	1	2	3	4
f_i	1.0	0.67	0.5	0.4	0.3

Estimate the area ABCD, using the trapezium rule.

2. Use the trapezium rule with

(i) $h = 1$, (ii) $h = 0.5$ and (iii) $h = 0.25$ to estimate $\displaystyle\int_0^1 \frac{1}{1+x}\,dx.$

NOTE   *Since we have found it convenient to set down the calculation of an estimate of a definite integral using the trapezium rule in tabular form, it can be inserted in a spreadsheet program very easily. Four columns of the spreadsheet can be used for values of x, f(x), 'Factor' and 'Product', as in the worked example and a fifth column can be used to form a cumulative total for the 'Product' column.*

Investigation

We want to investigate the accuracy obtained when the trapezium rule is used to evaluate a particular definite integral.

1. The trapezium rule is to be used to approximate $I = \displaystyle\int_{0.2}^{1.8} \sin(x+1)\,dx$,

using n strips. Using a computer package, or otherwise, complete the following table, where n is the number of strips used.

n	**Approximation to** I
2	
3	
4	
5	
6	

2. Given that $\displaystyle\int \sin(x+1)\,dx = -\cos(x+1) + C$, calculate the value of I

correct to 6 decimal places.

3. Add a further two columns to the table produced above to show the values of

(i)   the *absolute error*, $|\varepsilon| = |\text{exact value} - \text{approx value}|$

(ii) the *relative absolute error* $= \dfrac{|\varepsilon|}{|\text{exact value}|}$.

Comment on the results obtained.

The Error in the Trapezium Rule

From geometrical intuition and from the above results, it looks as though the approximation given by the trapezium rule can be improved by reducing the width of the strips; decreasing h increases the accuracy of the result. It would be useful to investigate how the error term depends on the value of h. Since, in each strip, the curve $y = f(x)$ has been approximated by a line, the error is represented by the shaded area in figure 3.5.

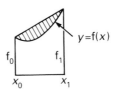

Figure 3.5

Use of the Taylor expansion of the function f gives a means of analysing the error term. To simplify the analysis, consider $\int_0^h f(x)\,dx$ with $f_0 = f(0)$, and $f_1 = f(h)$. Let I be the exact value of the integral and T be the approximation obtained using the trapezium rule.

Then
$$I = \int_0^h f(x)\,dx$$

$$= \int_0^h \left\{ f(0) + xf'(0) + \frac{x^2}{2!}f''(0) + \ldots \right\} dx,$$

using the Taylor expansion of f about $x = 0$

$$= \left[xf_0 + \frac{x^2}{2}f'(0) + \frac{x^3}{3!}f''(0) + \ldots \right]_0^h$$

$$= hf_0 + \frac{1}{2}h^2 f'(0) + \frac{1}{6}h^3 f''(0) + \ldots \qquad (1)$$

and
$$T = \frac{h}{2}(f_0 + f_1)$$

$$= \frac{h}{2}(f_0 + f(0 + h))$$

$$= \frac{h}{2}\left(f_0 + f(0) + hf'(0) + \frac{h^2}{2!}f''(0) + \ldots \right),$$

again, using the Taylor expansion of f about $x = 0$

$$= \frac{h}{2}\left(2f_0 + hf'(0) + \frac{1}{2}h^2 f''(0) + \ldots\right)$$

$$= hf_0 + \frac{1}{2}h^2 f'(0) + \frac{1}{4}h^3 f''(0) + \ldots \tag{2}$$

From (1) and (2),

$$I - T = \left(\frac{1}{6} - \frac{1}{4}\right)h^3 f''(0) + \text{higher degree terms in } h$$

$$= -\frac{1}{12}h^3 f''(0) + \text{higher degree terms.}$$

Hence, if h is small, a first approximation to the magnitude of the error inherent in using the trapezium rule is given by

$$|I - T| \approx \frac{1}{12}h^3 |f_0''|.$$

This is a *truncation error* since the error has resulted from the first degree approximation being used for f; that is, the Taylor expansion of f was cut short, or *truncated* to give the rule. The truncation error involves powers of h greater than 3 but the *principal error* term has magnitude $1/12\ h^3 |f_0''|$.

When the trapezium rule is applied over n strips with $nh = b - a$, the principal error in the *composite* formula is given by

$$|\text{error}| \approx \left|\frac{h^3}{12}f_0''\right| + \left|\frac{h^3}{12}f_1''\right| + \ldots + \left|\frac{h^3}{12}f_{n-1}''\right|$$

If the maximum value of $|f'(x)|$ for $x \in (a, b)$ is M then

$$|\text{error}| \leq n\frac{h^3}{12}M.$$

But $n = \dfrac{b-a}{h}$ so the principal error in the composite trapezium rule is

$$\frac{b-a}{h} \times \frac{h^3}{12}M = \frac{b-a}{12}h^2 M.$$

Since M is a property of the function f, and since the limits of integration, a and b, are given, the error can only be reduced by reducing the value of h.

Investigation

Investigate how well the principal truncation error formula predicts the actual error in applying the trapezium rule to a particular integral which can be evaluated analytically.

1. The integral to be used in this investigation is $I = \int_0^1 \sqrt{1+x}\,dx$

   Using calculus, obtain the value of this integral correct to 8 significant figures (or to the maximum number of figures given by your calculator).

2. An upper bound for the (principal) error in the composite trapezium rule is given by $\left|\dfrac{b-a}{12}M\right|h^2$ where $|f''(x)| < M$ for $x \in (a, b)$.

   Using a graph plotting program, or otherwise, obtain the value of $x \in (0, 1)$ at which $f''(x)$ takes it maximum value, $f(x)$ being given by $f(x) = \sqrt{(1+x)}$.

   Hence calculate

       (i)   the maximum value of $f''(x)$ for $x \in (0, 1)$,
   and (ii) an upper bound for the error, in terms of h.

3. Use a computer package to evaluate the approximations to

   $\int_0^1 \sqrt{1+x}\,dx$ using the trapezium rule, and complete the following

   table where

   n = the number of strips used
   h = the width of a strip
   T = the value obtained for the approximation to the integral
   I = value of integral obtained in task 1 above

n	h	T	$|T - I|$
10			
20			
30			
40			

   Does the expression for the error given in task 2 give an upper bound for the actual errors obtained in the above table?

4. Add a fifth column to the above table to verify that the magnitude of the error in the approximation to the integral is proportional to h^2.

It is important to realise that the formula for the principal error in the trapezium rule depends on the maximum value of the second derivative of f for $a \leq x \leq b$ which may be difficult to obtain. However, the formula does tell us that the error is proportional to h^2, written as $O(h^2)$, meaning that the error is *of the order of* h^2. This means that if the strip width is halved, the error will reduce by a factor of 4. In practice, the approximation is calculated using 2, 4, 8, etc. strips until a sufficiently accurate approximation is obtained.

Q

1. Use the trapezium rule with 1, 2, 4, 8, etc. strips to determine the value of $\int_{0}^{1} e^{-x^2} dx$, correct to 1 decimal place.

Let us return to the patio area problem described at the start of the chapter. The patio can be divided into strips 1 m wide as shown in figure 3.6 and the lengths of these strips measured to give the values shown in the table below.

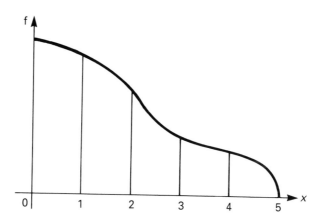

Figure 3.6

f_0	f_1	f_2	f_3	f_4	f_5
3	2.7	1.9	1.1	0.9	0.0

Using the trapezium rule,

$$\text{area} \approx \frac{1}{2}\{3 + 0.9 + 2(2.7 + 1.9 + 1.1 + 0.9)\}$$

$$\approx 8.1 \text{ m}^2$$

Since the concrete is to be 80 mm thick, the volume required is given by

$$\text{Volume} \approx 8.1 \times 0.08$$
$$\approx 0.648 \text{ m}^3.$$

In this problem, a high degree of accuracy was not required and the above estimate is probably adequate. However there are situations in which more accurate approximations are required.

This is true of many integrals which arise from 'simple' problems in dynamics and statistics. For example, if a simple pendulum oscillates with an angle of swing which cannot be considered as small, the period of the motion is given in terms of the integral

$$\int_a^b \frac{dx}{\sqrt{1 - k^2 \sin^2 x}}, \text{ where } k \text{ is a constant.}$$

Similarly, in statistics, the integral

$$\int_a^b e^{-t^2/2} \, dt$$

underlies the normal distribution which is one of the common statistical distributions. Neither of these integrals can be evaluated using any technique of integration available in calculus. However, graphs of the functions

$$f(x) = \frac{1}{\sqrt{1 - k^2 \sin^2 x}} \quad \text{and} \quad g(t) = e^{-t^2/2}$$

can be plotted and so the areas represented by the integrals over any interval (a, b) can be identified. Since the convergence of the trapezium rule can be slow, we should consider how the method might be modified to be more efficient. The method uses a geometrical approach, again the area required being approximated by a sum of component areas whose values *can* be calculated.

Simpson's Rule

The trapezium rule has the form

$$\int_a^b f(x) \, dx \approx \frac{h}{2} f_0 + h f_1 + h f_2 + \ldots + h f_{n-1} + \frac{h}{2} f_n$$

$$= \sum_{i=0}^{n} a_i f_i \quad \text{for some constants } a_i.$$

Some other rules of a similar form have also been derived. In *Simpson's Rule*, instead of fitting a line through the points (x_0, f_0) and (x_1, f_1) to form a trapezium, the strips are taken in *pairs*, as in figure 3.7, and a parabola is fitted through the three points (x_0, f_0), (x_1, f_1) and (x_2, f_2).

Figure 3.7

Let the parabola be

$$y = ax^2 + bx + c$$

and, to simplify the task, let A, M and B have x-values $-h$, 0 and h respectively so that D, N and C have coordinates $(-h, f_0)$, $(0, f_1)$ and (h, f_2), with $f_0 = f(-h)$, $f_1 = f(0)$ and $f_2 = f(h)$.

Then, if A is the area ABCD,

$$A = \int_{-h}^{h} (ax^2 + bx + c)\,dx$$

$$= \left[\frac{1}{3}ax^3 + \frac{1}{2}bx^2 + cx \right]_{-h}^{h}$$

$$= \frac{1}{3}ah^3 + \frac{1}{2}bh^2 + ch - \left(-\frac{1}{3}ah^3 + \frac{1}{2}bh^2 - ch \right)$$

$$= \frac{2}{3}ah^3 + 2ch.$$

To determine a, b and c for the parabola which passes through $(-h, f_0)$, $(0, f_1)$ and (h, f_2), these values are substituted in $y = ax^2 + bx + c$ to give

$$f_0 = ah^2 - bh + c$$
$$f_1 = 0 + 0 + c$$
$$f_2 = ah^2 + bh + c.$$

Solving these equations gives $f_1 = c$,

$$f_2 - f_0 = 2bh \text{ so that } b = \frac{1}{2h}(f_2 - f_0)$$

and $\qquad f_2 + f_0 = 2ah^2 + 2c \qquad$ so that $\quad 2ah^2 = f_2 + f_0 - 2f_1$

$$\text{giving} \qquad a = \frac{1}{2h^2}(f_2 + f_0 - 2f_1).$$

Substituting the expressions for a and c in the formula for A gives

$$A = \frac{2}{3} ah^3 + 2ch$$

$$= \frac{2}{3} \times \frac{1}{2h^2} (f_2 + f_0 - 2f_1)h^3 + 2hf_1$$

$$= \frac{h}{3} (f_2 + f_0 - 2f_1) + 2hf_1$$

$$= \frac{h}{3} (f_0 + 4f_1 + f_2)$$

So, for a pair of strips as shown in figure 3.7 where the curve DNC is a parabola, the area A is given by

$$A = \frac{h}{3} (f_0 + 4f_1 + f_2).$$

This result is known as Simpson's Rule.

HISTORICAL NOTE

Thomas Simpson (1710–1761) was a weaver from Spitalfields who taught himself mathematics and, as a break from working at his loom, taught mathematics to others. A textbook which he wrote in 1745 ran to 8 editions, the last of which was published in 1809. He became Professor of Mathematics at Woolwich College and was noted for his work on trigonometric proofs and for the derivation of formulae for use in the computation of tables of values of trigonometric functions. The result with which his name is associated had been published in draft form by the Scottish mathematician James Gregory in 1668 and was published in complete form by Simpson in his 'Mathematical Dissertation on Physical and Analytical Subjects' in 1743.

Over any interval (a, b) divided into an *even* number, $2n$, of strips, of width h, the *composite Simpson's rule* gives

$$\int_a^b f(x)\, dx \approx \frac{h}{3} \{f_0 + 4f_1 + f_2 + f_2 + 4f_3 + f_4 + \ldots + f_{2n-2} + 4f_{2n-1} + f_{2n}\}$$

$$\approx \frac{h}{3} \{f_0 + f_{2n} + 4(f_1 + f_3 + f_5 + \ldots + f_{2n-1}) + 2(f_2 + f_4 + \ldots + f_{2n-2})\}$$

$$\approx \frac{h}{3} \{f_0 + f_{2n} + 4 \text{ (sum of odd numbered ordinates)}$$
$$+ 2 \text{ (sum of even numbered ordinates)}\}$$

EXAMPLE   Use Simpson's rule with 4 strips to approximate $\int_1^2 \frac{1}{x}\,dx$.

Solution:

As with the trapezium rule, the solution should be set down in tabular form.

x	$f(x)$	Factor	Product
1.00	1.000	1	1.000
1.25	0.800	4	3.200
1.50	0.667	2	1.334
1.75	0.571	4	2.284
2.00	0.500	1	0.500
			8.318

$h = 0.25$ so, using Simpson's rule, $\int_1^2 \frac{1}{x}\,dx \approx \frac{0.25}{3} \times 8.318 = 0.693.$

Q

1. Use Simpson's rule with 4 strips to approximate $\int_1^2 \sqrt{1 + x^2}\,dx$.

2. Carrying four decimal places in your working, approximate $\int_0^1 \sin \pi x \,dx$ using Simpson's rule with

   (i) $h = \frac{1}{2}$ (ii) $h = \frac{1}{6}$ (iii) $h = \frac{1}{10}$

   Use calculus to obtain the value of the integral correct to 4 decimal places and calculate the absolute error in each of the approximations.

The Truncation Error in Simpson's Rule

To obtain the truncation error we can use a similar approach to the one used for the trapezium rule. In this case, we will work with a pair of strips as shown in figure 3.8.

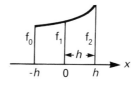

Figure 3.8

Let $f_0 = f(-h)$, $f_1 = f(0)$ and $f_2 = f(h)$.

The exact value of the integral I, is given by

$$I = \int_{-h}^{h} f(x)\,dx$$

$$= \int_{-h}^{h} \left\{ f(0) + xf'(0) + \frac{x^2}{2!}f''(0) + \frac{x^3}{3!}f'''(0) + \frac{x^4}{4!}f^{(iv)}(0) + \ldots \right\} dx$$

$$= \left[xf(0) + \frac{x^2}{2}f'(0) + \frac{x^3}{3!}f''(0) + \frac{x^4}{4!}f'''(0) + \frac{x^5}{5!}f^{(iv)}(0) + \ldots \right]_{-h}^{h}$$

$$= 2hf(0) + 2\frac{h^3}{3!}f''(0) + 2\frac{h^5}{5!}f^{(iv)}(0) + \ldots$$

$$= 2hf(0) + \frac{h^3}{3}f''(0) + \frac{h^5}{60}f^{(iv)}(0) + \ldots$$

Similarly the approximation to the integral obtained using Simpson's rule, S, is given by

$$S = \frac{h}{3}(f_0 + 4f_1 + f_2)$$

$$= \frac{h}{3}(f(-h) + 4f(0) + f(h))$$

$$= \frac{h}{3}\left[\left\{ f(0) - hf'(0) + \frac{h^2}{2!}f''(0) - \frac{h^3}{3!}f'''(0) + \frac{h^4}{4!}f^{(iv)}(0) + \ldots \right\} \right.$$

$$\left. + 4f(0) + \left\{ f(0) + hf'(0) + \frac{h^2}{2!}f''(0) + \frac{h^3}{3!}f'''(0) + \frac{h^4}{4!}f^{(iv)}(0) + \ldots \right\} \right]$$

$$= \frac{h}{3}\left[6f(0) + 2\frac{h^2}{2!}f''(0) + 2\frac{h^4}{4!}f^{(iv)}(0) + \ldots \right]$$

$$= 2hf(0) + \frac{h^3}{3}f''(0) + \frac{h^5}{36}f^{(iv)}(0) + \ldots$$

Note that the first two terms in the expressions for I and S agree and

$$I - S = \left(\frac{1}{60} - \frac{1}{36} \right) h^5 f^{(iv)}(0) + \ldots$$

$$= -\frac{1}{90} h^5 f^{(iv)}(0) + \text{terms of higher degree in } h.$$

So, over 1 pair of strips, the principal truncation error has magnitude

$$\frac{1}{90} h^5 |f^{(iv)}(0)|.$$

To obtain the error in the composite rule, over $2n$ strips with $b - a = 2nh$, the rule must be applied n times. If $|f^{(iv)}(x)| \leq M$ for x in the interval (a, b), an upper bound on the magnitude of the principal error term is given by

$$n \frac{h^5}{90} M = \frac{b-a}{2h} \times \frac{h^5}{90} M = \frac{b-a}{180} h^4 M.$$

Therefore the error in the composite Simpson's rule is of the order of h^4. So doubling the number of strips used in an application of the rule will reduce the error by a factor of 16.

NOTE *Since the error term involves fourth derivatives, Simpson's rule will give the exact value of the integral when applied to a cubic function.*

EXAMPLE By evaluating $\int_0^6 x^3 \, dx$ using Simpson's rule with 6 strips, verify that the rule gives the exact value of the integral.

Solution:

x	f(x)	Factor	Product
0	0	1	0
1	1	4	4
2	8	2	16
3	27	4	108
4	64	2	128
5	125	4	500
6	216	1	216
			972

$$h = 1, \quad \int_0^6 x^3 \, dx \approx \frac{1}{3} \times 972 = 324.$$

$$\text{Using calculus,} \quad \int_0^6 x^3 \, dx = \left[\frac{1}{4} x^4 \right]_0^6 = 324.$$

Investigation

Investigate how well the principal truncation error term predicts the actual error when Simpson's rule is used to estimate a definite integral.

1. The integral to be used in this investigation is

$$I = \int_0^1 \sin \pi x \, dx$$

First, using calculus, obtain the value of this integral correct to 7 decimal places.

2. An upper bound for the principal error in the composite Simpson's rule is given by

$$\frac{b - a}{180} M h^4$$

where

$$|f^{(iv)}(x)| \leq M \quad \text{for} \quad x \in (a, b).$$

Determine the value of $x \in (0, 1)$ at which $f^{(iv)}(x)$ takes it maximum value, where $f(x) = \sin \pi x$.

Hence calculate the maximum value of $f^{(iv)}(x)$ for $x \in (0, 1)$ and an upper bound for the error in terms of h.

3. Use a computer package to evaluate the approximations to the integral using Simpson's rule. Complete the following table where

n (an *even* number) = the number of strips used
h = the width of a strip
S = the value obtained for the approximation to the integral
I = the value of the integral obtained in **1**.

n	h	S	$|S - I|$
6			
10			
16			
20			
26			
30			

Does the expression for the error given in **2** give an upper bound for the actual errors obtained in the above table?

4. Add a fifth column to the above table to verify that the magnitude of the error in the approximation to the integral is proportional to h^4. Comment on how well the values fit this model of the error and, in particular, explain why you think the values fit the model less well for larger n.

Improving the Accuracy of the Integration Formulae

Although knowledge of the magnitude of the principal error term cannot be used directly to improve an approximation to an integral, we do know that the smaller the value of h, the better the approximation. However there is a danger in using too many strips. Each time an arithmetic operation is performed, there is likely to be a *round-off* error introduced. For example,

$$2 \div 3 = 0.66666666 \ldots$$

If a machine can carry only 5 figures, this will either be *rounded* to 0.66667 or *chopped* to 0.66666. Check what your calculator does. In either case, an error is introduced into the calculation. Although this error is small, if many thousand small errors of this type are introduced, their effect can become significant.

Just as it was useful to assess the magnitude of the truncation error in the integration formulae, so too is it useful to try to assess the effect of the round-off errors. In the trapezium rule, if we only consider round-off in the *final* step of the calculation of the function values, and if the maximum possible magnitude of each of these round-off errors is ε, then the magnitude of the maximum possible round-off error when n strips each of width h are used to approximate $\int_a^b f(x)\,dx$ is given by

$$\frac{h}{2}[\varepsilon + \varepsilon + 2(\varepsilon + \varepsilon + \varepsilon + \ldots \text{ to } (n-1) \text{ terms})]$$

$$= \frac{h}{2}[2\varepsilon + 2(n-1)\varepsilon]$$

$$= hn\varepsilon$$

$$= (b-a)\varepsilon.$$

You should verify that the same result holds for Simpson's rule. In each case, the error is independent of the number of strips. However if we looked more closely at errors which may occur *within* the computation of f_i we would find that, usually, the maximum possible round-off error will increase as the number of strips used is increased.

A simple technique has been developed for improving the accuracy of the approximation to the integral without increasing the number of strips used.

When the trapezium rule is used to approximate $\int_a^b f(x)\,dx$, using n

strips, each of width h, the truncation error ε_n is such that ε_n is approximately proportional to h^2. That is $\varepsilon_n \approx kh^2$ for some constant k.

Similarly if $2n$ strips are used, $\varepsilon_{2n} \approx k\left(\dfrac{h}{2}\right)^2 \approx \dfrac{1}{4}\varepsilon_n$.

Hence if I is the exact value of the integral, and T_n and T_{2n} are the approximations obtained using n and $2n$ strips,

$$I \approx T_n + \varepsilon_n \tag{1}$$

$$I \approx T_{2n} + \frac{1}{4}\varepsilon_n \tag{2}$$

Taking $[4 \times (2)] - (1)$ gives $\qquad 3I \approx 4T_{2n} - T_n$

$$\Rightarrow I \approx \frac{4}{3}T_{2n} - \frac{1}{3}T_n = T_{2n} + \frac{1}{3}(T_{2n} - T_n).$$

Thus the principal error term is greatly reduced if the better approximation T_{2n} is modified by an amount $1/3(T_{2n} - T_n)$, giving an improved approximation to I. It can be shown that the error in this improved approximation is of the order of h^4.

$$\varepsilon_n \propto h^4$$

$$\varepsilon_{2n} \propto \left(\frac{h}{2}\right)^4 \approx \frac{1}{16}\varepsilon_n$$

so that if S_n and S_{2n} are the approximations obtained using Simpson's rule,

$$I \approx S_n + \varepsilon_n \tag{3}$$

$$I \approx S_{2n} + \frac{1}{16}\varepsilon_n \tag{4}$$

Taking $[16 \times (4)] - (3)$ gives $\qquad 15I \approx 16S_{2n} - S_n$

$$\Rightarrow \quad I \approx \frac{16}{15}S_{2n} - \frac{1}{15}S_n$$

$$= S_{2n} + \frac{1}{15}(S_{2n} - S_n)$$

This technique is known as *Richardson's Extrapolation*.

HISTORICAL NOTE

Lewis Fry Richardson (1881–1953) was a scholar of the old school who made significant contributions in different fields of study. His early work was in physics but, an ardent pacifist, he became absorbed in problems of the psychology of peace and war and in 1929 he graduated in psychology. He also made significant contributions to the mathematics of weather prediction, developing techniques which only recently became practical with the introduction of high-speed computers. A native of the north of England, he served as principal of Paisley Technical College (now the University of Paisley) from 1929 till 1940 when the outbreak of war caused him to resign to pursue his peace studies in an idyllic corner of Argyll.

EXAMPLE

Determine approximations to $\int_0^{\pi/4} \dfrac{1}{\cos x}\,dx$ using the trapezium rule with 2 and 4 strips. Improve your approximation using Richardson's formula.

Solution:

x	$f(x)$	2 strips Factor	2 strips Product	4 strips Factor	4 strips Product
0.000	1.000	1	1.000	1	1.000
0.196	1.020			2	2.040
0.392	1.082	2	2.164	2	2.164
0.589	1.203			2	2.406
0.785	1.414	1	1.414	1	1.414
			4.578		9.024

For 2 strips, $h = 0.392$ giving

$$T_2 = \frac{0.392}{2} \times 4.578 = 0.897$$

For 4 strips, $h = 0.196$ giving

$$T_4 = \frac{0.196}{2} \times 9.024 = 0.884$$

Using Richardson's formula,

$$I \approx T_4 + \frac{1}{3}(T_4 - T_2)$$

$$= 0.884 + \frac{1}{3}(0.884 - 0.897)$$

$$= 0.880.$$

Since $T_4 = 0.884$ we can conclude that, to 2 decimal places, $I = 0.88$.

Q Carrying 6 decimal places in your working, determine approximations to

$$\int_0^{\pi/4} \frac{1}{\cos x} \, dx$$ using Simpson's rule with 2 and 4 strips. Improve your

approximation using Richardson's formula and quote your answer to the number of decimal places you expect to be correct.

Exercise 3

1. Use the trapezium rule to approximate the integral of the function f between $x = 10$ and $x = 16$, where values of f are given in the following table:

x	10	11	12	13	14	15	16
$f(x)$	10.1	4.5	5.4	10.7	14.7	16.6	17.5

2. Each of the following definite integrals can be obtained using calculus. Determine the integrals and compare the values with those obtained using the trapezium rule with the number of strips indicated

(i) $\displaystyle\int_1^3 \frac{dx}{x}; \, n = 4$ (ii) $\displaystyle\int_0^2 x^3 \, dx; \, n = 4$

(iii) $\displaystyle\int_0^1 \sin \pi x \, dx; \, n = 6$ (iv) $\displaystyle\int_0^{2\pi} x \sin x \, dx; \, n = 8$

3. Use the trapezium rule to evaluate the following integrals using 2, 4 and 8 strips. State the value of the integral to the maximum number of figures you consider to be correct.

(i) $\displaystyle\int_{0.5}^1 x^3 e^x \, dx$ (ii) $\displaystyle\int_1^3 \frac{x}{\sqrt{1 + x^2}} \, dx$

4. Repeat questions 1 to 3, using Simpson's rule.

5. A function f has values given in the following table.

x	1.8	2.0	2.2	2.4	2.6
$f(x)$	3.12014	4.42569	6.04241	8.03014	10.46675

Approximate $\displaystyle\int_{1.8}^{2.6} f(x) \, dx$ using (i) the trapezium rule and (ii) Simpson's rule.

Exercise 3 continued

If the data are found to contain errors as given in the following table, estimate the resulting errors in the approximations to the integral.

x	1.8	2.0	2.2	2.4	2.6
Error in $f(x)$	2×10^{-6}	-2×10^{-6}	-0.9×10^{-6}	-0.9×10^{-6}	2×10^{-6}

6. Approximate $\displaystyle\int_{2}^{4} \frac{\ln x}{x}\, dx$ using Simpson's rule with (a) 4 strips

(b) 8 strips. Hence use Richardson's method to improve the approximations.

7. Obtain approximations to the value of $\displaystyle\int_{1}^{2} \ln x\, dx$ using both the

trapezium and Simpson's rules, using 2 strips.

In each case, use the formula for the principal truncation error to calculate an upper bound on the error term.

Given that $\displaystyle\int_{1}^{2} \ln x\, dx = 0.38629$, correct to 5 decimal places, comment

on the actual errors obtained.

8. In practice, the actual truncation error is usually much less than the upper limit given by the error formula.

To illustrate this, repeat question 7, using the integral $\displaystyle\int_{0}^{0.1} x^{1/3}\, dx$.

Extended Questions

1. The value of π is to be estimated using numerical integration. Two methods are proposed:

(i) to estimate the area in the first quadrant of the circle

$$x^2 + y^2 = 1.$$

and (ii) to estimate the area in the shaded sector of the circle, shown overleaf, using strips parallel to Oy, between the line OP and the circumference of the circle.

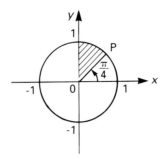

Using the same number of strips in each case, which method gives the better result?

By examining the truncation error terms in the formulae used, explain why one method is superior to the other.

2. Derive an integration formula based on approximating the area of the strip from x_0 to $x_0 + h$ by the area of the rectangle of width h and height

   (i) $f(x_0)$ and   (ii) $f(x_0 + 1/2h)$.

   Determine estimates of the truncation error in each formula, comparing the results with those obtained for the trapezium rule and for Simpson's rule. In each case, derive the corresponding Richardson extrapolation formula.

3. Let T_n and T_{2n} be estimates of the value of a definite integral,

   $$I = \int_a^b f(x)\,dx,$$ obtained using the trapezium rule with n and $2n$ strips.

   Richardson's method gives an improved estimate of I which could be written as

   $$T_{n,2n} = T_{2n} + \frac{1}{3}(T_{2n} - T_n).$$

   Putting $n = 1$, verify that $T_{n,2n}$ gives the Simpson's rule approximation to the integral with two strips.

   It can be shown that

   $$I = T_n + c_1 h^2 + c_2 h^4 + c_3 h^6 + \ldots$$

   where the quantities of c_i, $i = 1, 2, 3, \ldots$, are functions of the derivatives of f and are independent of h. Use this result to show that an approximation to I with a truncation error of the order of h^6 is given by

   $$T_{n,2n,4n} = T_{2n,4n} + \frac{1}{15}(T_{2n,4n} - T_{n,2n})$$

Extended Questions continued

These approximations can be set down as follows

$$T_n$$
$$\downarrow$$
$$T_{2n} \rightarrow T_{n,2n}$$
$$\downarrow$$
$$T_{4n} \rightarrow T_{2n,4n} \rightarrow T_{n,2n,4n}$$
$$\downarrow$$

producing successively better approximations to I. This procedure is known as *Romberg integration*.

Obtain the formula for $T_{n,2n,4n,8n}$ and hence evaluate $\displaystyle\int_0^1 \frac{dx}{2+x^4}$,

(to five decimal places) starting with a strip of width 1.

KEY POINTS

- The composite form of the trapezium rule, using n strips each of width h gives

$$\int_a^b f(x)\,dx \approx \frac{h}{2}(f_0 + f_n + 2[f_1 + f_2 + \ldots + f_{n-1}])$$

- The principal term in the truncation error in this formula has magnitude no greater than $b - a/12\, h^2 M$ where M is the maximum value of $|f''(x)|$ for $x \in (a, b)$.
- The composite form of Simpson's Rule, using $2n$ strips each of width h gives

$$\int_a^b f(x)\,dx \approx \frac{h}{3}(f_0 + f_{2n} + 4[f_1 + f_3 + \ldots + f_{2n-1}]) + 2[f_2 + f_4 + \ldots + f_{2n-2}])$$

- The principal term in the truncation error in this formula has magnitude no greater than $b - a/180\, h^4 M$ where M is the maximum value of $|f^{(iv)}(x)|$ for $x \in (a, b)$.

- If T_n and T_{2n} are the approximations to an integral obtained using the trapezium rule with n and $2n$ strips, respectively, a better approximation to the integral is given by

$$T_{2n} + 1/3(T_{2n} - T_n)$$

- The corresponding result for Simpson's rule is that an improved approximation is given by

$$S_{2n} + 1/15(S_{2n} - S_n)$$

4

The Solution of Differential Equations

As an introduction to the work in this chapter, consider the solution of the following two related problems.

Problem 1

A student borrows £500 to purchase a motorcycle and is charged interest at the rate of 2% per month. Assuming that he has paid nothing back, how much does he owe after 3 months?

Solution:

Let s_n be the sum owing after n months.

Then
$$s_{n+1} = s_n + 0.02s_n$$
$$= 1.02s_n, \qquad s_0 = 500$$

Hence
$$s_1 = 1.02 \times 500 \qquad = 510$$
$$s_2 = 1.02 \times 510 \qquad = 520.20$$
$$s_3 = 1.02 \times 520.20 \ = 530.60$$

Problem 2

On a trip to the coast, the motorcycle tyre develops a puncture. The tyre loses air more slowly as the pressure decreases and

$$\frac{dP}{dt} = -0.001P$$

If the student pumps the tyre up to 40 psi, what is the pressure of the air in the tyre after 3 seconds?

Before considering the solution of Problem 2, we note that the two problems are similar; just as the student's overdraft changes with time, so also does the air pressure in the tyre of the motorcycle. However, as illustrated in figure 4.1, whereas the debt changes only once a month, the air pressure is changing continuously.

The equation

$$\frac{dP}{dt} = -0.001P$$

is called a *differential equation*. Some differential equations, including this one, can be solved using calculus. You may be able to solve the equation and you should verify that the solution is

$$P = 40e^{-0.001t}$$

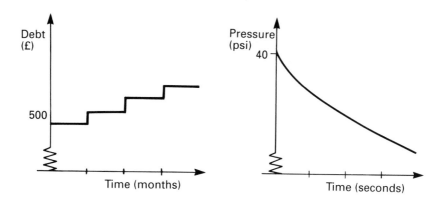

Figure 4.1

For many differential equations, a solution cannot be obtained using calculus. However it is sometimes possible to use an approach similar to that used in the overdraft problem to approximate the solution of a differential equation. We will investigate such a method, using the differential equation

$$\frac{dP}{dt} = -0.001P, \quad P = 40 \text{ when } t = 0$$

as an illustration.

The first step will be to replace the rate of change of pressure, $\dfrac{dP}{dt}$, by a numerical approximation to the derivative. This process is called *numerical differentiation* and we will look at two possible approximations in this chapter.

Numerical Differentiation

Consider a function, f, whose graph is shown in figure 4.2.

The slope of the curve $y = f(x)$ at A where $x = x_0$ is given by $f'(x_0)$ and, for small values of h, this could be approximated by the slope of the chord AB where B is the point $(x_0 + h, f(x_0 + h))$.

Therefore $\qquad f'(x_0) \approx \dfrac{1}{h} \{f(x_0 + h) - f(x_0)\}.$

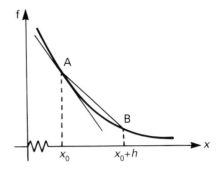

Figure 4.2

This approximation to $f'(x_0)$ uses points on the curve which lie to the right of $x = x_0$ and is called a *forward difference* formula. Taylor's theorem can be used to give an estimate of the error in this formula as follows.

Since $\qquad f(x_0 + h) = f(x_0) + hf'(x_0) + \dfrac{h^2}{2!} f''(\eta)$ where $x_0 < \eta < x_0 + h$,

$$f(x_0 + h) - f(x_0) = h\{f'(x_0) + \frac{h}{2!} f''(\eta)\}$$

$$\Rightarrow \qquad \frac{1}{h}\{f(x_0 + h) - f(x_0)\} = f'(x_0) + \frac{h}{2!} f''(\eta)$$

Therefore $\qquad f'(x_0) = \dfrac{1}{h} \{f(x_0 + h) - f(x_0)\} - \dfrac{h}{2} f''(\eta)$

It follows that the *truncation error* introduced when the forward difference formula is used to approximate $f'(x_0)$ is of the order of h, that is $O(h)$. From work done in Chapter 3 on the integration formulae, we know that very small values of h are required to give acceptable results when the truncation error is so large. Looking again at figure 4.2 we realise that a better approximation to $f'(x_0)$ might be given by using function values on both the left *and* the right of the point A.

For example, in figure 4.3, if P is the point $(x_0 - h, f(x_0 - h))$ and Q is the point $(x_0 + h, f(x_0 + h))$, the slope of the chord PQ could be used to approximate the slope of the tangent to the curve at $x = x_0$, giving

$$f'(x_0) \approx \frac{1}{2h} \{f(x_0 + h) - f(x_0 - h)\}.$$

This is an example of a *central difference formula*.

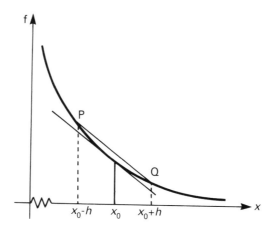

Figure 4.3

Again, Taylor's theorem can be used to examine the order of convergence of this approximation.

Since

$$f(x_0 + h) = f(x_0) + hf'(x_0) + \frac{h^2}{2!} f''(x_0) + \frac{h^3}{3!} f'''(\eta_1)$$

where $x_0 < \eta_1 < x_0 + h$

and

$$f(x_0 - h) = f(x_0) - hf'(x_0) + \frac{h^2}{2!} f''(x_0) - \frac{h^3}{3!} f'''(\eta_2)$$

where $x_0 - h < \eta_2 < x,$

$$f(x_0 + h) - f(x_0 - h) = 2hf'(x_0) + \frac{h^3}{3!} \{f'''(\eta_1) + f'''(\eta_2)\}$$

$$= 2h \left[f'(x_0) + \frac{h^2}{12} \{f'''(\eta_1) + f'''(\eta_2)\} \right]$$

$$\Rightarrow \quad \frac{1}{2h} \{f(x_0 + h) - f(x_0 - h)\} = f'(x_0) + \frac{h^2}{12} \{f'''(\eta_1) + f'''(\eta_2)\}$$

So

$$f'(x_0) = \frac{1}{2h} \{f(x_0 + h) - f(x_0 - h)\} + O(h^2).$$

This shows that the central difference approximation will, in general, give a better approximation to $f'(x_0)$ than that given by the forward difference formula.

EXAMPLE

Given that $f(x) = 1/x$, obtain forward difference and central difference approximations to $f'(1.1)$, taking h as (i) 0.1, (ii) 0.05 and (iii) 0.01.

Compare the results you obtain with the exact value.

Solution:

Since $f(x) = \dfrac{1}{x}$, $f'(x) = -\dfrac{1}{x^2}$ and $f'(1.1) = -0.826$ to 3 decimal places.

Using a forward difference formula,

$$f'(1.1) \approx \frac{1}{h}\{f(1.1+h) - f(1.1)\}$$

This gives the results shown in the following table.

h	0.1	0.05	0.01
$f'(1.1)$	−0.758	−0.791	−0.819
|error|	0.068	0.035	0.007

Using a central difference formula,

$$f'(1.1) \approx \frac{1}{2h}\{f(1.1+h) - f(1.1-h)\}$$

This gives the results below.

h	0.1	0.05	0.01
$f'(1.1)$	−0.833	−0.828	−0.8265
|error|	0.007	0.002	0.0005

Q

Using the same values of h as in the above example, and the same function f, obtain approximations to $f'(1.5)$, comparing your results with the exact value.

From pure mathematics, we know that

$$f'(x_0) = \lim_{h \to 0} \frac{f(x_0 + h) - f(x_0)}{h}$$

and so the smaller the value used for h in the forward difference formula, the better should be the approximation obtained. The previous examples confirm this. However, the following investigation will indicate that the situation is not so straightforward.

Investigation

Investigate how the accuracy of the numerical differentiation formulae depends on the value of h, for some given functions, f.

1. The function, f, is given by $f(x) = x^3$. The computation should be carried out on a computer where the following BASIC program could be used.

```
10 X = 1
20 H = 1
30 F = X^3
40 FOR K = 1 TO 10
50 F1 = (X + H)^3
60 PRINT H,"    ",(F1-F)/H
70 H = H/10
80 NEXT K
```

   Running this program will print out values of the step size h and the forward difference approximation to $f'(1)$ where $f(x) = x^3$ and $h = 1$, $0.1, 0.01 \ldots, 10^{-9}$. Examine the results obtained and compare these with the exact value of $f'(1)$.

2. Modify the above program to print values of h and the corresponding approximations to $f'(1)$ given by the central difference formula. Experiment with values of x other than $x = 1$ and summarise your results.

3. Repeat 1 and 2 using $f(x) = \sin x$ with $x = 1.5$. Comment on the results you obtain.

You will have observed that, as h decreases, the approximations to $f'(x)$ approach the value of $f'(x)$ for a number of steps; then one or two 'wrong' values appear followed by zero values for the remaining approximations. Also, the breakdown in the sequence of approximations occurs at about the same value of h in each case.

To detect what is going wrong, it is useful to work through an example by hand.

Let $f(x) = \ln x$. We want to examine approximations to $f'(2.2)$. Suppose that we record the values of the logarithms used and all calculated values to 3 decimal places. Taking $h = 0.04, 0.004, 0.0004$ and 0.00004 gives the following results:

$$h = 0.04 \quad , \quad f'(2.2) \approx \frac{\ln 2.24 - \ln 2.2}{0.04} = \frac{0.806 - 0.788}{0.04} = 0.450$$

$$h = 0.004 \quad , \quad f'(2.2) \approx \frac{\ln 2.204 - \ln 2.2}{0.004} = \frac{0.790 - 0.788}{0.004} = 0.500$$

$$h = 0.0004 \quad , \qquad f'(2.2) \approx \frac{\ln 2.2004 - \ln 2.2}{0.0004} = \frac{0.789 - 0.788}{0.0004} = 2.500$$

$$h = 0.00004 \quad , \qquad f'(2.2) \approx \frac{\ln 2.20004 - \ln 2.2}{0.00004} = \frac{0.788 - 0.788}{0.00004} = 0$$

Note that $f'(x) = \dfrac{1}{x}$ so that $f'(2.2) = 0.455$, correct to 3 decimal places.

It becomes clear that the problem is rooted in the decision to round off to 3 decimal places. Working to this accuracy, for all $h < 0.00004$, $\ln(2.2 + h) = \ln 2.2$. Although the computer is working to greater accuracy, a similar problem arises, leading to the results found in the investigation.

It is interesting to look more closely at the calculation of the approximation to $f'(2.2)$ with $h = 0.0004$. This gave the spurious value, $f'(2.2) \approx 2.5$.

If we carry 6 decimal places in the calculation,

$$\ln 2.2004 = 0.788639$$
and
$$\ln 2.2 \quad\; = 0.788457$$
so that
$$\ln(2.2004) - \ln(2.2) = 0.000182.$$

In the calculation to 3 decimal places this was taken as $0.789 - 0.788$ which is equal to 0.001. Clearly an error in the 4th decimal place in the value of $\ln(2.2)$ is relatively *insignificant* but an error in the 4th decimal place of $\ln(2.2004) - \ln(2.2)$ has serious consequences.

Great care must be taken when subtracting nearly equal quantities to avoid the effects of this increase in the *significance* of errors. In the numerical differentiation problem, the position is made considerably worse by dividing $\ln(2.2004) - \ln(2.2)$ by the small number 0.0004, that is, multiplying by $1/0.0004$ or 2500.

From these considerations, we see that, although pure mathematics suggests that h should be made very small, the limitations of computer technology require that an optimum value of h be used; for this value, the truncation error arising is small, but the value of h is not so small that the round-off errors in the computation become significant.

It is indeed the case that using numerical differentiation formulae is fraught with problems! However, there are situations where they are used and we will proceed to use the forward difference approximation to the derivative to provide a numerical solution to the differential equation of Problem 2 at the start of the chapter.

So from Problem 2,

$$\frac{dP}{dt} = -0.001P \quad \text{with} \quad P = 40 \text{ when } t = 0.$$

First replace $\dfrac{dP}{dt}$, at a particular time t, by the forward difference approximation,

$$\frac{dP}{dt} \approx \frac{P(t+h) - P(t)}{h}$$

If we take $h = 1$ and let P_r, $r = 0, 1, 2$, etc. be the pressure in the tyre after r seconds then $P_0 = 40$ and

since
$$\frac{P(0+1) - P(0)}{1} \approx -0.001P(0)$$

$$\frac{P_1 - P_0}{1} \approx -0.001P_0$$

$\Rightarrow$
$$P_1 \approx P_0 - 0.001P_0 = 0.999P_0 = 39.960$$

Similarly,
$$\frac{P_2 - P_1}{1} \approx -0.001P_1$$

$\Rightarrow$
$$P_2 \approx P_1 - 0.001P_1 = 0.999P_1 = 39.920$$

and
$$P_3 \approx P_2 - 0.001P_2 = 0.999P_2 = 39.880.$$

Proceeding in this way, after r seconds,

$$\begin{aligned}
P_r &\approx 0.999P_{r-1} \\
&= (0.999)^2 P_{r-2} \\
&= (0.999)^r P_0 \\
&= (0.999)^r \times 40.
\end{aligned}$$

We noted earlier that the exact solution of this equation is
$$P = 40e^{-0.001t}$$

Q Construct a table showing the exact and approximate solutions of the differential equation discussed above for $t = 20\,\text{s}$, $40\,\text{s}$, $60\,\text{s}$, $80\,\text{s}$ and $100\,\text{s}$.

The above result shows excellent agreement between the solutions; although such good results are not common, the following method, known as *Euler's method* is of interest.

Euler's Method

HISTORICAL
NOTE

Leonhard Euler was one of the leading mathematicians of the 18th century. The son of a Swiss clergyman, he followed his father's wishes and studied theology and Hebrew. However, while at the University of Basle, he took private lessons in mathematics and his father was persuaded to allow him to follow a career in mathematics.

His contributions in the fields of mathematics and physics were wide ranging, including work on calculus, geometry, celestial mechanics and the theory of numbers. One particularly useful contribution was to introduce much of the mathematical notation now taken for granted, including π, e, Σ, $\sin x$, A, B and C for the angles of a triangle and a, b and c for the sides of a triangle.

He fathered 13 children of whom 5 survived to adulthood. From the age of 60 he began to lose his eyesight but his mathematical productivity did not diminish. He had a phenomenal memory and remarkable computational powers; with the assistance of two of his sons who wrote down the results of his work, he continued active in research till his death in 1783 at the age of 76.

Generalising the method used to solve Problem 2, we will consider the solution of the differential equation

$$\frac{dy}{dx} = f(x, y) \text{ with } y = y_0 \text{ when } x = x_0.$$

To solve this equation we must find a curve $y = y(x)$ which passes through the point (x_0, y_0) and is such that the slope at any point (x, y) on the curve is given by $f(x, y)$. Replacing the derivative $\frac{dy}{dx}$ in the equation by its forward difference approximation at $x = x_0$ gives

$$\frac{1}{h}\{y(x_0 + h) - y(x_0)\} + \frac{h}{2} = f(x_0, y_0), \qquad \text{where } x_0 < \eta < x_0 + h$$

Rearranging this equation to obtain $y(x_0 + h)$ gives

$$y(x_0 + h) - y(x_0) + -\frac{h^2}{2}y''(n) = hf(x_0, y_0)$$

$$\Rightarrow \quad y(x_0 + h) \approx y(x_0) + hf(x_0, y_0)$$

This gives an approximation to the solution at $x = x_0 + h$ the approximation containing a truncation error which is of the order of h^2.

Putting $\qquad x_n = x_0 + nh$ and $y_n = y(x_n)$ we have

$$y_1 \approx y_0 + hf(x_0, y_0),$$
$$y_2 \approx y_1 + hf(x_1, y_1) \text{ etc.}$$

In general, $\qquad\qquad y_{n+1} \approx y_n + hf(x_n, y_n).$

This formula describes *Euler's method* for approximating the solution of the given differential equation. Figure 4.4 shows the exact solution, $y = y(x)$, and the set of points y_n, $n = 1, 2, 3, \ldots$ obtained using Euler's method.

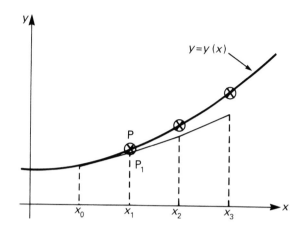

Figure 4.4

Note that, in Euler's method, the curve $y = y(x)$ between $x = x_0$ and $x = x_1$ is approximated by the line through (x_0, y_0) and of slope $f(x_0, y_0)$. The truncation error in the approximation of $y(x_1)$ is represented by the distance between the points $P(x_1, y(x_1))$ and $P_1(x_1, y_1)$ in figure 4.4.

EXAMPLE   Given the differential equation

$$\frac{dy}{dx} = x + y \text{ with } y(0) = 1,$$

use Euler's method to obtain approximations to the solution at $x = 0.02$, 0.04 and 0.06. Carry three decimal places in the calculation.

Solution:
Euler's method gives
$$y_{n+1} = y_n + hf(x_n, y_n)$$
For the given equation,
$$f(x, y) = x + y \quad \text{so}$$
$$y_{n+1} = y_n + h(x_n + y_n),$$
$$x_0 = 0, \ y_0 = 1, \ h = 0.02.$$

Hence
$$y_1 = y_0 + h(x_0 + y_0)$$
$$= 1 + 0.02(0 + 1)$$
$$= 1.02 \qquad \text{with } x_1 = 0.02$$
$$y_2 = y_1 + h(x_1 + y_1)$$
$$= 1.02 + 0.021$$
$$= 1.041 \qquad \text{with } x_2 = 0.04$$
$$y_3 = y_2 + h(x_2 + y_2)$$
$$= 1.041 + 0.02(0.04 + 1.041)$$
$$= 1.063 \qquad \text{with } x_3 = 0.06$$

This could be set down in tabular form as follows:

x_n	y_n	$f(x_n, y_n)$ $= x_n + y_n$	$hf(x_n, y_n)$ $= h(x_n + y_n)$	y_{n+1} $= y_n + h(x_n + y_n)$
0.00	1.000	1.000	0.020	1.020
0.02	1.020	1.040	0.021	1.041
0.04	1.041	1.081	0.022	1.063
0.06	1.063			

Q

1. Use Euler's method to solve the equation given in the above example at $x = 0.01, 0.02, 0.03$ and 0.04.

2. Use Euler's method to obtain approximations to the solution of the differential equation

$$\frac{dy}{dx} = x - y + 1,$$

with $y(0) = 1$. Use a step size $h = 0.1$ and continue the solution to $x = 0.5$.
Let A_i be the point (x_i, y_i) for $i = 0, 1, 2, 3, 4, 5$.
Show the points A_i on a graph, joining the points with line segments.
It can be shown that the exact solution of this equation is $y = x + e^{-x}$.
Calculate the exact solution at $x = x_i$ for $i = 1, 2, 3, 4$ and 5 and calculate the errors in the numerical solution at each point.

The Truncation Error in Euler's Method

In the derivation of Euler's method, we showed that

$$y_1 = y_0 + hf(x_0, y_0) + O(h^2)$$

That is, the truncation error in *one step* of the process is of the order of h^2. This error is called the *local truncation* error. The process is a *step by step method* and if it is applied over n steps, the *global truncation error* can be taken as approximately proportional to nh^2. However n is dependent on h, with n steps of length h lying between x_0 and x_n, so that

$$nh = x_n - x_0 \text{ and hence } nh^2 = (x_n - x_0)h.$$

It follows that the global truncation error is of the order h. This is a low order of error and, in general, could only give a good approximation to the function y if a large number of steps are used, making h small.

The diagram in figure 4.5 illustrates the build-up of errors from step to step; if the exact solution is $y = y(x)$ and P is the point (x_0, y_0), the tangent at P is drawn to give the point Q (x_1, y_1).

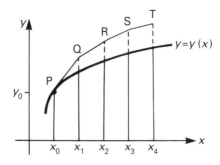

Figure 4.5

The slope of the tangent to $y = y(x)$ at $x = x_1$ is approximated and used to give the line QR; R is the point on this line at which $x = x_2$.

The slope of $y = y(x)$ at $x = x_2$ is now approximated and the process repeated.

Clearly the error is carried from one step to the next and a new error is introduced at each step, leading to the global error being of the order h.

Investigation

Investigate the effect on the solution of a given differential equation of reducing the step size in Euler's method. An equation which *can* be solved analytically is used so that exact and numerical solutions can be compared.

1. Consider the differential equation

$$\frac{dy}{dx} = 2x \text{ with } y = 0 \text{ when } x = 0.$$

Using your calculator, apply Euler's method with a step size of length 1 to complete the following table

x	0	1	2	3	4	5	6	7	8
y	0								

2. Use calculus to solve the equation and draw a graph to display the solution and the set of points obtained above on the same axes.

Investigation continued

3. Use a computer to solve the equation with a step size of 0.5 by
   either (i) using a spreadsheet program to set up the x_i values 0.0, 0.5,
   1.0, 1.5, . . ., 8.0 in column A and the y_i values,
   $y_{i+1} = y_i + 0.5 \times 2 \times x_i$, in column B
   or (ii) using the following BASIC program to obtain $y(x)$ for $x = 0.0$
   to 8.0.

```
10 X = 0
20 Y = 0
30 H = 0.5
40 FOR X = 0 TO 8 STEP H
50 Y = Y + 2*H*X
60 PRINT X,"   ",Y
70 NEXT X
```

   Show the values obtained at $x = 0.0$ to 8.0 on the graph produced in **2**.

4. Use the method in 3 to obtain the solution with a step size of 0.2.
   Compare the results.

5. Repeat **2** and **3** for the equation

$$\frac{dy}{dx} = 2x \text{ with } y = 2 \text{ when } x = 0.$$

   Comment on the results obtained.

The Modification of Euler's Method

Since the global truncation error associated with Euler's method is of the order h, reliable solutions will be obtained only if h is very small and a large number of steps is used. Since using a large number of steps increases the round-off error, it is advisable to look for a method for which the truncation error is of higher order.

Consider the equation,

$$\frac{dy}{dx} = f(x, y) \text{ with } y = y_0 \text{ when } x = x_0.$$

The solution of this equation is $y = y(x)$ and, using the same notation as used in Euler's method, integrating the differential equation with respect to x, between the limits $x = x_0$ and $x = x_1$ gives

$$\int_{x_0}^{x_1} \frac{dy}{dx} dx = \int_{x_0}^{x_1} f(x, y) \, dx$$

$$\Rightarrow \quad \left[y\right]_{x_0}^{x_1} = \int_{x_0}^{x_1} f(x, y)\, dx$$

$$\Rightarrow \quad y_1 - y_0 = \int_{x_0}^{x_1} f(x, y)\, dx$$

$$\Rightarrow \quad y_1 = y_0 + \int_{x_0}^{x_1} f(x, y)\, dx$$

As a first approximation, we could assume that $f(x, y) = f(x_0, y_0) = f_0$ for all x in the interval $x_0 \le x \le x_1$ so that

$$y_1 \approx y_0 + f_0 \int_{x_0}^{x_1} dx$$

$$= y_0 + f_0(x_1 - x_0)$$

$$= y_0 + hf_0, \text{ which is Euler's formula.}$$

Using a better approximation to $\int_{x_0}^{x_1} f(x, y)\, dx$ should give a more accurate solution to the differential equation. Using the trapezium rule we have

$$y_1 \approx y_0 + \int_{x_0}^{x_1} f(x, y)\, dx$$

$$= y_0 + \frac{h}{2}(f_0 + f_1) + O(h^3) \tag{1}$$

the error term, $O(h^3)$ being produced when the trapezium rule is applied to a single strip.

However, $f_1 = f(x_1, y_1)$ and the value of y_1 is the solution of the differential equation at $x = x_1$ which is the value we are trying to determine! One possible approach is to use Euler's method to *approximate* y_1 and to use the formula based on the trapezium rule to improve this approximation.

So approximating y_1 using Euler's method gives

$$y_1 \approx y_0 + hf_0$$

Using (1), an improved approximation to y_1 is given by

$$y_1 \approx y_0 + \frac{h}{2}(f_0 + f_1) \tag{2}$$

$$= y_0 + \frac{h}{2}\{f(x_0, y_0) + f(x_1, y_1)\}$$

$$\approx y_0 + \frac{h}{2}\{f(x_0, y_0) + f(x_1, y_0 + hf_0)\}$$

To illustrate this modified Euler method, consider the equation

$$\frac{dy}{dx} = 2(x+y)\cos x \text{ with } y = 1 \text{ when } x = 0.$$

Taking $h = 0.2$ and using Euler's method to obtain y_1 we obtain

$$
\begin{aligned}
y_1 &\approx y_0 + hf(x_0, y_0) \text{ with } f(x, y) = 2(x+y)\cos x \\
&= 1 + 0.2 \times 2(0+1)\cos 0 \\
&= 1.4
\end{aligned}
$$

The exact solution $y = y(x)$ and this Euler approximation to $y(0.2)$ are represented in figure 4.6.

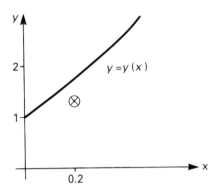

Figure 4.6

Now $f(x_0, y_0) = f(0, 1) = 2$ and using the approximation to $y(0.2)$, we can approximate $f(x_1, y_1)$ by

$$f(0.2, 1.4) = 2 \times 1.6 \cos 0.2 = 3.1362 \text{ to 4 decimal places}$$

Therefore from equation (2) which uses the trapezium rule

approximation to $\displaystyle\int_0^{0.2} f(x, y)\, dx$ as illustrated in figure 4.7, we have

$$y_1 \approx y_0 + \frac{h}{2}(f_0 + f_1)$$

$$= 1 + 0.1(2 + 3.1362)$$

$$= 1.5136$$

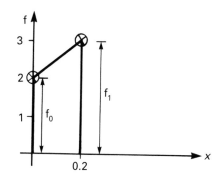

Figure 4.7

A Runge-Kutta Method

In each of the step-by-step methods studied, we obtained an increment k such that

$$y_1 = y_0 + k$$

which gives us an approximation to $y(x_1)$. With Euler's method applied

to the equation $\dfrac{dy}{dx} = f(x, y)$, $k = hf(x_0, y_0)$, as illustrated in figure 4.8.

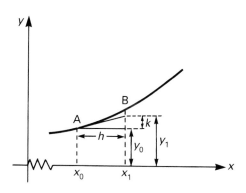

Figure 4.8 Euler's method

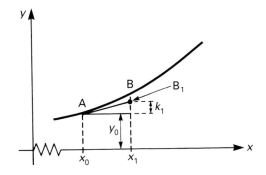

Figure 4.9 Modified Euler method

To obtain the value of k for the modified Euler method, let $k_1 = hf(x_0, y_0)$ so that the point B_1 in figure 4.9 has coordinates $(x_0 + h, y_0 + k_1)$. Then $f(x_0 + h, y_0 + k_1)$ will give an approximation to the slope of the solution curve at B. The mean of the approximations to the slopes at A and B, namely,

$$\frac{1}{2}\{f(x_0, y_0) + f(x_0 + h, y_0 + k_1)\}$$

will approximate the slope of AB so that a better approximation to B will be given by

$$y_1 = y_0 + \frac{1}{2}h\{f(x_0, y_0) + f(x_0 + h, y_0 + k_1)\}.$$

Letting

$$k_2 = hf(x_0 + h, y_0 + k_1)$$

gives

$$y_1 = y_0 + \frac{1}{2}(k_1 + k_2).$$

Therefore in general, given a point (x_0, y_0) on the solution curve, we define

$$k_1 = hf(x_n, y_n)$$

and

$$k_2 = hf(x_n + h, y_n + k_1)$$

so that the modified Euler approximation to y_{n+1} is given by

$$y_{n+1} = y_n + \frac{1}{2}(k_1 + k_2) \qquad \text{for } n = 0, 1, 2, \ldots$$

When the method is written in this form, in terms of k_1 and k_2, it is one of a class of methods known as Runge-Kutta methods. This particular method is a second order Runge-Kutta method, since the local truncation error can be shown to be of the order h^2.

Carl Runge was an outstanding mathematician and experimental physicist who worked in Germany in the early 1900s. He was noted for his computational skills and when the Wright brothers made their first flight, he constructed a model of their plane with paper, weighted with needles. Using this he was able to give a good estimate of the capacity of the motor used, the details of which were a closely guarded secret. The method for solving differential equations named after him was published in 1895 and modified by Kutta in 1901.

In practice, when determining numerical solutions of differential equations using these methods, the equation is solved initially with a step length, h, then with a step length, $\frac{h}{2}$, and the solutions compared to give an estimate of the accuracy being achieved.

EXAMPLE

Use the Runge-Kutta method to solve the equation

$$\frac{dy}{dx} = x + y, \quad y(0) = 1,$$

at $x = 0.02$ and 0.04 using a step of length 0.02. Repeat the calculation with a step of length 0.01 and hence write down the values of y at $x = 0.02$ and $x = 0.04$, as accurately as your calculations will permit.

Solution:
The Runge-Kutta formulae are
$$k_1 = hf(x_n, y_n)$$
$$k_2 = hf(x_n + h, y_n + k_1)$$

$$y_{n+1} = y_n + \frac{1}{2}(k_1 + k_2)$$

In this example, $x_0 = 0$, $y_0 = 1$, $f(x, y) = (x + y)$.

Taking $h = 0.02$,
$$k_1 = 0.02(0 + 1) = 0.02$$
$$k_2 = 0.02(0.02 + 1.02) = 0.0208$$

At $x_1 = 0.02$, $\quad y_1 = 1 + 0.5(0.02 + 0.0208) = \underline{1.020400}$

Continuing the process,
$$k_1 = 0.02(0.02 + 1.0204) = 0.020808$$
$$k_2 = 0.02(0.04 + 1.041208) = 0.021624$$

At $x_2 = 0.04$, $\quad y_2 = \underline{1.041616}$

Reducing the step size to $h = 0.01$,

$$k_1 = 0.01(0 + 1) = 0.01,$$
$$k_2 = 0.01(0.01 + 1.01) = 0.0102$$

At $x_1 = 0.01$,
Continuing,
$$y_1 = 1 + 0.5(0.0202) = \underline{1.0101}$$
$$k_1 = 0.01(0.01 + 1.0101) = 0.010201$$
$$k_2 = 0.01(0.02 + 1.020301) = 0.010403$$

At $x_2 = 0.02$,

$y_2 = 1.0101 + 0.5(0.020604)$
$= \underline{1.020402}$

Continuing,

$k_1 = 0.01(1.040402) = 0.010404$
$k_2 = 0.01(0.03 + 1.030806) = 0.010608$

At $x_3 = 0.03$,

$y_3 = 1.020402 + 0.5(0.021012)$
$= \underline{1.030908}$

Continuing,

$k_1 = 0.01(1.060908) = 0.010609$
$k_2 = 0.01(0.04 + 1.041517) = 0.010815$

At $x_4 = 0.04$,

$y_4 = 1.030908 + 0.5(0.021424)$
$= \underline{1.041620}$

Examining these results shows that, at $x = 0.02$, using 1 step of width 0.02 gave $y_1 = 1.020400$ and using 2 steps of width 0.01 gave $y_2 = 1.020402$ suggesting that, correct to 5 decimal places,

$$y(0.02) = 1.02040$$

The corresponding values at $x = 0.04$ are

$$y_2 = 1.041616 \text{ and } y_4 = 1.041620$$

giving $\qquad y(0.04) = 1.04162$, correct to 5 decimal places.

The equation $\dfrac{dy}{dx} = \dfrac{-y^2}{1 + x}$, $y(0) = 1$

is to be solved at $x = 0.1$. Use the Runge-Kutta method with (i) $h = 0.1$ and (ii) $h = 0.05$ to obtain an approximation to the solution. Quote your answer to the maximum number of figures you consider to be correct.

Exercise 4

1. Given that $f(x) = \cos x$, obtain forward and central difference approximations to $f'(0)$ and $f'(1.5)$, taking (i) $h = 0.2$, (ii) $h = 0.04$ and (iii) $h = 0.008$. Compare the values obtained with the values of $f'(0)$ and $f'(1.5)$ respectively. Explain why the central difference formula performs so well in approximating $f'(0)$.

2. The following five points lie on a curve $y = f(x)$:

$$(0, -1), (0.25, -0.9), (0.5, -1), (0.75, -1.1), (1, -1).$$

Taking the step size as 0.25, obtain forward and central difference approximations to $f'(0.5)$.
Obtain, also, approximations to $f'(0.25)$ and $f'(0.75)$ and hence approximate $f''(0.5)$ using both forward and central differences.
Comment on the values obtained.
Given that the function f can be approximated by the polynomial

$$p(x) = 2x^3 - 3x^2 + x - 1$$

Exercise 4 continued

comment on the approximations obtained for the derivatives. In
particular, by considering p"(0.5) explain why the forward and central
difference formula give the same results for f'(0.5).
Calculate values of p(x) for $x = 0.48, 0.49, 0.50, 0.51$ and 0.52 and use
these values to estimate f'(0.5) and f"(0.5). How good are the
approximations in this case?

3. Use the forward difference approximation to f'(x_0), namely,

$$f'(x_0) \approx \frac{f(x_0 + h) - f(x_0)}{h}$$

to obtain an approximation to f"(x_0) in terms of $f(x_0 + 2h)$, $f(x_0 + h)$ and
$f(x_0)$.

4. Use Euler's method to solve the following differential equations at
$x = p$, using the starting values and step sizes as specified.

(i) $\dfrac{dy}{dx} = \dfrac{x}{y}$, $y(0) = 1, h = 0.1, p = 0.4$

(ii) $\dfrac{dy}{dx} = 1 + 2xy$, $y(0) = 1, h = 0.1, p = 0.4$

(iii) $\dfrac{dy}{dx} = 2y + \sin x$, $y(0) = 2, h = 0.1, p = 0.4$

(iv) $\dfrac{dy}{dx} = \dfrac{(\sin x) - y}{x}$, $y(1) = -0.540, h = 0.1, p = 1.3$

(v) $\dfrac{dy}{dx} = x^3 - 2xy$, $y(1) = 0, h = 0.1, p = 1.4$

(vi) $2y\dfrac{dy}{dx} = 1 - \sin x$, $y(0) = 1, h = 0.1, p = 0.4$

5. The differential equation

$$\frac{dy}{dx} = x + y + xy, \ y(0) = 1$$

is to be solved at $x = 0.1$.
Use the Runge-Kutta method with $h = 0.025$ to obtain an
approximation to the solution giving your answer to 4 decimal places.

6. Use the Runge-Kutta method to obtain approximate solutions at
$x = 0.1(0.1)0.5$ for the equation

$$\frac{dy}{dx} = \sin x + y, \ y(0) = 2.$$

Exercise 4 continued

7. Use the Runge-Kutta method to solve the equation

$$10 - 5\frac{dy}{dx} = y, \ y(0) = 0, \ h = 0.5$$

at $x = 1, 2$, and 3.
The exact solution is given by

$$y = 10(1 - e^{-0.2x})$$

Write down the exact solution at $x = 1, 2$ and 3 and calculate the % error in each of the estimated values.

Extended Questions

1. Given a function $y = y(x)$, write down the Taylor polynomial of degree 1 which approximates $y(x_0 + h)$.
Show that this approximation could be used to derive Euler's method for solving the differential equation

$$\frac{dy}{dx} = f(x, y) \text{ with } y = y_0 \text{ when } x = x_0,$$

giving the local truncation error as $O(h^2)$.
Use Taylor polynomials to derive an approximation to $y(x_0 + h)$ which has $O(h^2)$ truncation error of the order of h^4.
Use this approximation to solve

$$\frac{dy}{dx} = 2y + x^2 e^x, \text{ with } y = 0 \text{ when } x = 1,$$

at $x = 2$, taking $h = 0.1$.
Use this equation to compare the performance of Euler's method and the fourth order Taylor method.

2. Given the differential equation $\frac{dy}{dx} = f(x, y), \ y = y_0$ when $x = x_0$

a Runge-Kutta method which has local truncation error of the order h^4 is given by

$$y_1 = y_0 + \frac{1}{6}(k_1 + 2k_2 + 2k_3 + k_4)$$

where $k_1 = hf(x_0, y_0)$

$$k_2 = hf\left(x_0 + \frac{h}{2}, y_0 + \frac{k_1}{2}\right)$$

$$k_3 = hf\left(x_0 + \frac{h}{2}, y_0 + \frac{k_2}{2}\right)$$

$$k_4 = hf(x_0 + h, y_0 + k_3)$$

Exercise 4 continued

Show that, for the special case of the equation $\dfrac{dy}{dx} = F(x)$, with

(x_0, y_0) given, this method can be derived by approximating

$\displaystyle\int_{x_0}^{x_0+h} F(x)\,dx$ using the Simpson's rule approximation to the integral.

Use the equation

$$\frac{dy}{dx} = \left(\frac{y}{x}\right)^2 + \left(\frac{y}{x}\right) \quad \text{with } y = 1 \text{ when } x = 1,$$

where the solution is required at $x = 1.2$ correct to 4 decimal places, to compare the performance of the given second and fourth order Runge-Kutta methods.

KEY POINTS

- Taylor's Theorem can be used to give approximations to the first derivative of a function f at a point $x = x_0$.
- The *forward difference* approximation is

$$f'(x_0) = \frac{1}{h}\{f(x_0 + h) - f(x_0)\} - O(h^2)$$

- A *central difference* approximation is

$$f'(x_0) = \frac{1}{2h}\{f(x_0 + h) - f(x_0 - h)\} - \frac{h^2}{6}f'''(\eta)$$

where $x_0 < \eta < x_0 + h$.

- Two *step-by-step* method to approximate the solution of

$$\frac{dy}{dx} = f(x, y) \quad \text{with } y(x_0) \text{ given,}$$

(i) *Euler's Method*

$$y_{n+1} = y_n + hf(x_n, y_n)$$

with error term $\dfrac{1}{2}h^2 f'(\eta)$ where $x_n < \eta < x_n + h$;

when the method is applied to one step, the *local truncation error* is of the order of h^2 but when the method is applied over a number of steps, the *global truncation error* has order h.

(ii) *Runge-Kutta Method*

$$k_1 = hf(x_n, y_n) \; ; \; k_2 = hf(x_n + h, y_n + k_1)$$

and

$$y_{n+1} = y_n + \frac{1}{2}(k_1 + k_2)$$

5

Errors in Numerical Processes

So far we have used mathematical concepts to develop numerical algorithms for solving a collection of mathematical problems. In some cases, methods have had to be modified to take account of errors which are either inherent in the process used or inevitably introduced in the computation. In this final chapter, a number of strands which have been running through the course are drawn together and the experience gained in handling numerical methods is used to heighten our awareness of the difficulties which can arise in producing numerical solutions to problems. It is important to remember that, in general, using a numerical algorithm to solve the mathematical model of a physical problem will be only a part of the complete solution. The solution could consist of several steps which are summarised in figure 5.1.

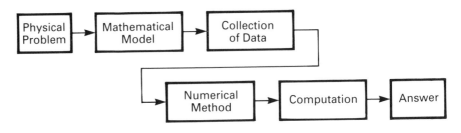

Figure 5.1

As an illustration of these steps consider the example of a simple pendulum oscillating as shown in figure 5.2. The displacement of the pendulum in the x-direction is required at a given time t.

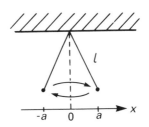

Figure 5.2

This problem can be tackled as follows.

Step 1
The analyst must look at the problem and make any approximations he considers valid to obtain a simpler problem which he can handle. For example, he may decide to consider only the effect of gravity on the motion and ignore any air resistance. In this way, a *mathematical model* of the situation is constructed. Because of the approximations made, a modelling error, ε_p, is now present in the process. The relationship produced may involve the length of the string, l, the maximum displacement of the bob, a, etc. These measurements must be made in the next step.

Step 2
If the data to be collected involve natural numbers they are said to be *discrete* and can be expressed exactly. For example:
(i)   the number of live births per 1000 of the population in Greater Glasgow in one year
(ii) the number of cars crossing the Severn Bridge per hour.

However, if the data are expressed in terms of real numbers, as in the pendulum example where measurements of distance have to be made, the data can only be as accurate as the measuring device will permit. Therefore a measurement error ε_m is introduced.

Step 3
When the data values have been obtained and inserted in the mathematical model, the problem becomes one of numerical analysis and the best method of solution must be chosen. As seen in the methods studied in the course, numerical methods usually require approximations to be made; for example, in an iterative method, the iteration must be stopped after a finite number of steps. Another error ε_n will usually be introduced at this stage.

Step 4
Having chosen the method, the solution must now be computed either by hand or using a machine. In either case, rounding will take place, introducing further errors ε_r.

The question which must be asked now is, having made errors $\varepsilon_p + \varepsilon_m + \varepsilon_n + \varepsilon_r$ on the way through the process, does the number obtained at the end give a solution to the original physical problem?

Of the four errors described, the numerical analyst must be aware of the existence of those made at Steps 1 and 2 but they are beyond his control. Attempts must be made to minimise those introduced at Step 3 by choosing a method which converges fast or which introduces as small a truncation error as is feasible. We must also minimise those introduced in Step 4 by reducing the effects of rounding errors in the computation.

Ideally, we would like to be able to give an upper bound for the magnitude of the truncation error and for the magnitude of the round-off errors but this is not always possible.

In this chapter, we review the errors introduced in the numerical solution of a problem and examine some simple techniques which can be used to minimise their effect. We must be sure that numerical results are produced efficiently and that they are reliable. Earlier chapters looked at efficiency; this chapter is concerned with reliability.

Truncation Errors in Numerical Processes

Most of the numerical algorithms which have been studied are based on infinite processes which are truncated. For example

(i)   When solving an equation $f(x) = 0$ using an iterative formula

$$x_{r+1} = g(x_r), \text{ with } x_0 \text{ given,}$$

the required root is the limit of the sequence x_0, x_1, x_2 etc.; some criterion is used to decide when to terminate this sequence.

(ii) When Taylor polynomials are used to approximate the value of a function at a point $x_0 + h$, for example,

$$f(x_0 + h) \approx f(x_0) + hf'(x_0) + \frac{h^2}{2!} f''(x_0),$$

a truncation error is introduced.

(iii) The numerical integration formulae are derived from truncated infinite series and so have inherent truncation errors.

In some cases, it is possible to derive an upper bound for the truncation error but, frequently, this is not easy to compute.

A problem can arise in deciding when to terminate an iterative process: In Chapter 1, the simple rule used was that when two successive iterates are the same to p figures, we assume that this value gives the limit of the sequence correct to p figures. This is valid when the sequence is converging fast. For example, the sequence

$$u_{r+1} = u_r + \frac{1}{3^r}, \text{ with } u_1 = 1$$

gives the values shown in the following table.

r	1	2	3	4	5	6	7
u_r	1	1.3333	1.4444	1.4815	1.4938	1.4979	1.4993

The last two values in the table agree to 2 decimal places and we would assume that $\lim_{r \to \infty} u_r = 1.50$, correct to 2 decimal places.

Since

$$u_1 = 1$$

$$u_2 = 1 + \frac{1}{3}$$

$$u_3 = 1 + \frac{1}{3} + \frac{1}{3^2}$$

$$u_4 = 1 + \frac{1}{3} + \frac{1}{3^2} + \frac{1}{3^3}$$

we see that the sequence will converge to the sum to infinity of the geometric series

$$1 + \frac{1}{3} + \frac{1}{3^2} + \frac{1}{3^3} + \ldots$$

which is 1.5.

However, for the sequence $u_{r+1} = u_r + 1/r$, $u_1 = 1$, the successive values of u_r are changing slowly. When the values are printed out, we see that

$$u_{140} = u_{141} = 6.52, \text{ correct to 2 decimal places.}$$

But we would be wrong to assume that the limit of u^r is 6.52, since looking at values of u_r for $r > 141$, we see that u_r approaches ∞ as r approaches ∞.

Another problem with a slowly converging series is that round-off errors are occurring at each step so that, after a large number of steps, they may become significant. So to avoid stopping an iteration prematurely and to avoid build-up of round-off errors, it is important when choosing an iterative process that we choose one which converges fast.

$\boxed{Q}$ It can be shown that

$$\sin x = x - \frac{x^3}{3!} + \frac{x^5}{5!} - \frac{x^7}{7!} + \ldots$$

Use this series to approximate $\sin 0.1$ correct to 4 significant figures. How many terms must be taken so that the truncation error in the approximation is less than 10^{-6}? Obtain an expression for the principal truncation error when $\frac{\sin x}{x}$ is approximated by 1, for small x.

The Measurement of Errors

Before considering the sources and the effects of errors caused by rounding numbers, we should consider how errors are measured. Clearly, the error, ε, when a given value x is approximated by x^*, is given by

$$\varepsilon = x - x^*$$

This measure gives no information about the significance of the error; for example, an error of 0.1 m in a measurement of a window frame could have serious consequences, but an error of 0.1 m in the measurement of the distance by road from London to Bristol would be trivial. A more useful measure is the *relative error* defined by

$$\text{relative error} = \frac{\text{error}}{\text{exact value}} = \frac{\varepsilon}{x}$$

That is, by measuring the error as a fraction of the value itself.

In handling numerical errors, we are not usually concerned about whether the error is positive or negative but the *magnitude* of the error is important. For this reason the *absolute error*, defined as $|\varepsilon|$, and the

absolute relative error, $\left|\dfrac{\varepsilon}{x}\right|$, are used.

Recording Numbers

Numbers can be written in either *fixed point* form in which the decimal point is in position

e.g.   123.45

or in *floating point* form (also called scientific notation) in which the number is expressed as $a \times 10^b$ (or $a\,\mathrm{E}b$) with $1 \le |a| < 10$ and $b \in \mathbb{N}$,

e.g. 1.2345×10^2

Very large and very small numbers are most conveniently expressed in floating point form. Floating point form is also more informative about the accuracy of the value represented. Consider the situation in which a metre-long piece of wood is cut into three equal pieces; knowledge of the saw used suggests that the length of each piece should be taken as 0.33 m. This could be written as 330 mm or 3.3×10^2 mm. From the first of these values, without further information, we might expect the measurement to have been exactly 330 mm which we know to be unlikely. A more accurate fixed point representation of the measurement would be 330 ± 5 mm. The floating point form gives only the digits which are correct; these are the *significant* figures in the measurement. In the fixed point form the two threes are significant but the zero is indicating the place value of the other digits. Had the measurement of 330 mm been exact, it would have been represented in floating point form by 3.30×10^2 mm, with 3 significant digits.

Write the numbers (i) 1.230×10^{-3}, (ii) 3.140×10^5 and (iii) 1.02100×10^5 in fixed point form. How many significant digits does each number contain?

Solution:
(i) $1.230 \times 10^{-3} = 0.001230$ with 4 significant figures. Note that the zero is inserted at the end of the decimal part of the fixed point representation to indicate that 4 figures are significant.
(ii) $3.140 \times 10^5 = 314000$ with 4 significant figures.
(iii) $1.02100 \times 10^5 = 102100$ with 6 significant figures.

Q

1. Given that the fixed point number 21500 contains only 4 significant figures, write the number in floating point form.

2. Express the number (i) 1.40×10^{-2} and (ii) 7.200×10^4 in fixed point form. How many significant figures does each number contain?

Rounding Errors

Whether we are doing calculations by hand or by machine, there will be a maximum number of digits which can be used in a number. For example,

$$2 \div 3 = 0.6666 \ldots$$

In calculations the decimal part of this number must be terminated. Some computers *chop* the decimal part by neglecting all digits after the last one to be carried. Chopping the above value so that it contains only 5 digits after the decimal point would give 0.66666.

However, if only 5 decimal digits can be carried, it is more usual to round the given number to the *nearest* number with 5 decimal digits, namely 0.66667. Note that the maximum possible error introduced by rounding a number to 5 decimal places is $0.000005 = 5 \times 10^{-6}$. In general, the magnitude of the maximum rounding error when working to p decimal places is $5 \times 10^{-p-1}$.

Similarly, if the floating point number 2.34566×10^8 is rounded to 5 significant figures, we obtain 2.3457×10^8 and the magnitude of the rounding error is 0.00004×10^8 or $4 \times 10^{-5} \times 10^8$. In general, if $a \times 10^b$ is rounded to s significant figures, the maximum possible rounding error would be $5 \times 10^{-s} \times 10^b$.

Note that when $a \times 10^b$ is rounded to s significant figures,

$$|\text{relative error}| \leq \left| \frac{5 \times 10^{-s} \times 10^b}{a \times 10^b} \right|$$

$$\leq \left| \frac{5 \times 10^{-s} \times 10^b}{10^b} \right| \quad \text{since } a \geq 1$$

$$\leq 5 \times 10^{-s}$$

This argument can be taken in reverse and, if the relative error in a number is less than or equal to 5×10^{-s} then the number is correct to s significant figures.

EXAMPLE

Calculate the absolute error and the absolute relative error in approximating

(i)   0.00246 to 4 decimal places

(ii) 401.555 to 4 significant figures.

Solution:

(i) Rounding to 4 decimal places gives 0.0025

Hence $\qquad |\text{error}| = |0.00246 - 0.0025| = 0.00004$

and $\qquad |\text{relative error}| = \dfrac{0.00004}{0.00246} = 0.016$ to 2 significant figures.

(ii) $\qquad\qquad\qquad 401.555 = 4.01555 \times 10^2$

$\qquad\qquad\qquad\qquad\quad = 4.016 \times 10^2$ to 4 significant figures

$\qquad\qquad\qquad\qquad\quad = 401.6$

Hence $\qquad |\text{error}| = |401.555 - 401.6| = 0.045$

and $\qquad |\text{relative error}| = \dfrac{0.045}{401.555} = 0.00011$ to 2 significant figures.

Q

1. Round the number 203.664 to (i) 4 significant figures and (ii) 2 decimal places. Calculate the absolute error and the absolute relative error in each approximation.

2. Write down the maximum possible absolute error when a number is rounded to 4 decimal places. What is the maximum possible relative error when the number is rounded to 4 significant figures?

   Round the number 1/300 to (i) 4 decimal places and (ii) 4 significant figures. Calculate the absolute error in (i) and the absolute relative error in (ii).

The use of computers has made possible the solution of very large problems. A computer may work 10^9 times faster than a human and, although it does not make 'mistakes' as a human would, rounding takes place at all stages of the computation. The following exercises are chosen to highlight the limitations of the calculating devices we use and should be worked through using a calculator or computer. The results obtained will depend on the equipment used and any unexpected results should be recorded and explained in your solutions.

Q

1. Use a calculator or computer to evaluate

   (i) $12000 - 0.00012$
   (ii) $120000 - 0.000012$
   (iii) $1200000 - 0.0000012$

2. Enter the number 123456 in your calculator; multiply by 9 repeatedly until the display changes to floating point form. What is the largest number held in fixed point form? Similarly, find the smallest (positive) number which can be held in fixed point form by entering 0.1 and dividing by 9 repeatedly.

3. Key in the following BASIC program and run it to determine the smallest number which can be added to 1.

```
10  E = 1
20  REPEAT
30  E = E/2
40  PRINT E
50  UNTIL 1 + E = 1
```

   Modify this program to determine the 'next' number after (i) 100, (ii) 1000. Carry out a similar exercise on your calculator.

Investigation

What is the round-off error in computer arithmetic in a computation involving summation?

1. In algebra, the *associative law* states that $a + (b + c) = (a + b) + c$. This result is not always true in computation where storage requirements make rounding off necessary. For example, using floating point arithmetic with 4 significant figures evaluate each of the expressions $a + (b + c)$ and $(a + b) + c$ where $a = 1.234$, $b = 0.0002$ and $c = 0.0004$. That is, $a = 1.234$, $b = 2.000 \times 10^{-4}$ and $c = 4.000 \times 10^{-4}$. Write down your results.

2. Write down the values of s and $s - 1$ where $s = (0.01 + 0.01 + 0.01 + \ldots$ to 100 terms). The following BASIC program calculates s and prints the values of $s - 1$.

Investigation continued

```
10  S = 0
20  FOR K = 1 TO 100
30  S = S + 0.01
40  NEXT K
50  PRINT S-1
```

Type in this program and run it, noting the result.
Modify the program to print out the value of $s - 1$ where

(i) $s = (0.001 + 0.001 + \ldots$ to 100 terms)

and (ii) $s = (0.0001 + 0.0001 + \ldots$ to 10000 terms)

Calculate the absolute error in each of the results and comment on its magnitude as the number of terms increases.

Propagation of Errors

Although it is fairly easy to put an upper bound on the error or relative error incurred in rounding a number, it is much more difficult to calculate a bound on the error in the answer to a computation which used rounded values. For example, if

$$A = \frac{1}{2}ab \sin C$$

and a and b have been rounded to 1 decimal place and the angle C is in radians, rounded to 3 decimal places, how accurate is the value of A? We will not attempt to find an answer to such a question, but some guidelines are derived which might help in making an intelligent guess at the number of figures which might be reliable in the answer. We will look at three particular calculations.

1. Addition (and subtraction) calculations

Consider the situation in which two lengths, l_1 and l_2, are to be added. If $l_1 = 3.33 \times 10^2$ mm and $l_2 = 1.342 \times 10$ mm, we know that $l_1 = 333$ mm with a possible error of ± 0.5 mm and $l_2 = 13.42$ mm with a possible error of ± 0.005 mm.

Then $l_1 + l_2 = 333 + 13.42 = 346.42$ mm.

Since the error in l_1 is ± 0.5 mm, the value of $l_1 + l_2$ is unlikely to be accurate to 2 decimal places and the value of $l_1 + l_2$ should be given as 346 mm. This gives us the first guideline.

Rule 1:   The number of decimal places quoted in a sum should be no more than the number quoted in the term with the least number of decimal places.

For example, $1.23 + 12.14 + 0.567 = 13.937$. The first two terms have possible errors of 0.005 and so the sum may contain an error of 0.005. So, only two figures should be quoted giving 13.94. Looking at the possible errors in more detail, the maximum possible error is of magnitude $0.005 + 0.005 + 0.0005$ which equals 0.0105, so that the value 13.94 is probably not even accurate to 2 decimal places.

2. Multiplication (and division) calculations

Treating the general case, if x and y are exact values which are approximated by x^* and y^* respectively, then

$$xy = (x^* + \varepsilon_x)(y^* + \varepsilon_y) \quad \text{where } \varepsilon_x \text{ and } \varepsilon_y \text{ are the errors in } x \text{ and } y \text{ respectively.}$$

$$= x^* y^* + x^* \varepsilon_y + y^* \varepsilon_x + \varepsilon_x \varepsilon_y$$

$$\Rightarrow \qquad xy - x^* y^* = x^* \varepsilon_y + y^* \varepsilon_x + \varepsilon_x \varepsilon_y$$

$$\Rightarrow \qquad \frac{xy - x^* y^*}{xy} = \frac{x^* \varepsilon_y}{xy} + \frac{y^* \varepsilon_x}{xy} + \frac{\varepsilon_x \varepsilon_y}{xy}$$

If it is assumed that ε_x and ε_y are small so that the product $\varepsilon_x \varepsilon_y$ is negligible and if it is also assumed that $\dfrac{x^*}{x} \approx \dfrac{y^*}{y} \approx 1$

then

$$\frac{xy - x^* y^*}{xy} \approx \frac{\varepsilon_y}{y} + \frac{\varepsilon_x}{x}$$

Therefore, the relative error in $xy \approx$ the relative error in $x +$ the relative error in y.

It was shown that a value is correct to p significant figures if and only if the relative error in the value is less than 5×10^{-p}. Suppose that x^* is correct to p significant figures and y^* is correct to q significant figures then

$$\left| \frac{\varepsilon_x}{x} \right| \leq 5 \times 10^{-p} \text{ and } \left| \frac{\varepsilon_y}{y} \right| \leq 5 \times 10^{-q}$$

and so

$$\text{absolute relative error in } xy = \left| \frac{xy - x^* y^*}{xy} \right|$$

$$\leq 5 \times 10^{-p} + 5 \times 10^{-q}$$

If $p > q$ then $10^{-p} < 10^{-q}$ and so the relative error in xy can probably be approximated by 5×10^{-q}. Hence the number of significant figures in xy will be no more than the number of significant figures in the less accurate data value. This gives the second guideline.

Rule 2: The number of significant figures in a product (or quotient) should be no more than the number of significant figures in the factor with least significant figures.

For example, the rule suggests that

$$1.23456 \times 2.3$$

should only be quoted to 2 significant figures giving 2.8. Checking this by using the largest and smallest possible values of the data gives

$$1.234555 \times 2.25 \leq \text{product} \leq 1.234565 \times 2.35$$
$$\Rightarrow \qquad 2.7777488 \leq \text{product} \leq 2.9012278$$

In this example, the product could be outside the interval 2.75 to 2.85 but the rule does give a reasonable indication of the number of figures to quote in the product.

3. Subtraction of Nearly Equal Quantities

This is perhaps the most important of the three guidelines in the sense that if it is inadvertently not adhered to, large errors may result.

Consider the evaluation of

$$98.357(10.234 \times 12.063 - 123.420)$$

Applying Rules 1 and 2 to the contents of the brackets suggest, at a casual glance, that it should be possible to evaluate the expression correct to 5 significant figures to give 3.2204. However, the calculation simplifies to give

$$\begin{aligned} \text{Expression} &= 98.357(123.453 - 123.420) \\ &= 98.357 \times 0.033 \\ &= 3.2 \text{ correct to only 2 significant figures.} \end{aligned}$$

This gives us the third guideline.

Rule 3: Whenever two positive numbers of nearly equal magnitude are subtracted, there may be a loss of *many* significant figures in the accuracy of the answer.

If the calculation were done by hand, such a loss of significant figures would be obvious; the danger arises when the calculation is performed on a calculator or computer and the intermediate values are not obvious.

EXAMPLE   Use Rules 1 and 2 to evaluate the following assuming that all values are correct only to the number of figures shown.

(i) $0.0032 + 1.29$ (ii) $129.85 - 114.367$

(iii) 0.0032×1.29 (iv) $55.37(19.273 - 1.94)$

Solution:

(i) $0.0032 + 1.29 = 1.2932$

Since 1.29 was quoted to only 2 decimal places, so the sum can only be quoted to 2 decimal places to give 1.29.

(ii) $129.85 - 114.367 = 15.48$ (to 2 decimal places)

(iii) $0.0032 \times 1.29 = 0.004128$

Since the first factor contains only 2 significant figures so the product should be quoted as 0.0041.

(iv) $55.37(19.273 - 1.94) = 55.37 \times 17.33$ (2 decimal places)

$$= 959.6 \text{ (4 significant figures)}.$$

Ill-conditioned Problems

In this final section, we look at some problems which are particularly sensitive to small errors. These problems are said to be ill-conditioned and are characterised by the property that a small change in the information input causes a large change in the output as illustrated in figure 5.3.

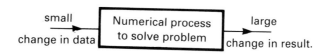

Figure 5.3

Examples of such problems are

(i) $x^2 - 2x + 0.9\ = 0$

which has 2 solutions

but $x^2 - 2x + 1.01 = 0$

has no solution

(ii)
$$\left. \begin{array}{c} x - y = 1 \\ x - 1.0001y = 0 \end{array} \right\}$$ has solution $x = 10001$

$y = 10000$

$$\left. \begin{array}{c} \text{but } x - y = 1 \\ x - 0.9999y = 0 \end{array} \right\}$$ has solution $x = -9999$

$y = -10000$

Since the coefficients in these equations could be data values or could have been obtained from a previous computation and so be subject to errors, the problem of obtaining reliable solutions to the equations is difficult to solve. Looking at the two examples, in the first, if $f(x) = x^2 - 2x + 0.9$ we see from the graph of the function f in figure 5.4 that f has a turning-point very close to the x-axis.

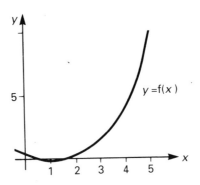

Figure 5.4

Similarly, looking at the equations in (ii), the first can be written as $y = x - 1$ and the second as $y = 0.9999x$ which are represented geometrically by the pair of almost parallel lines, shown in figure 5.5.

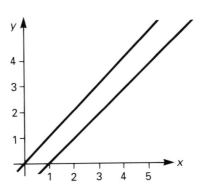

Figure 5.5

In this section, some other ill-conditioned problems are examined, the aim being to make you more familiar with such difficulties. There is no simple rule for handling ill-conditioned problems except the obvious one that in all work associated with the problem, extra figures should be carried to minimise the effects of round-off error.

Examine the values of $f(x) = \dfrac{1}{1 - x^2}$ for values of x near to 1.

Solution:

The values are listed in the following table.

x	$f(x)$	Change in x	Change in f
0.9	5.26316		
		0.09	44.9881
0.99	50.2513		
		0.009	449.999
0.999	500.250		
		0.0009	4500
0.9999	5000.25		

Thus, a change of 0.0009 in the x value, from $x = 0.999$ to $x = 0.9999$ gives a change of 4500 in the calculated value of $f(x)$, displaying serious ill-conditioning. Note that, in this case, the subtraction of nearly equal numbers is responsible for the ill-conditioned nature of the problem.

Q Construct a similar table showing values of $f(x) = \dfrac{1}{1 - x^2}$ for

(i) $x = 1.1, 1.01, 1.001$ and 1.0001 and (ii) $x = 6.1, 6.01, 6.001$ and 6.0001. Is the evaluation of $f(x)$ ill-conditioned in either of these cases?

It may be interesting to examine the graph of the above function f to see if it contains any clues to assist in finding the cause of the problem.

The function $y = f(x)$ is shown in figure 5.6. We can see that $x = 1$ and $x = -1$ are *asymptotes* and near to these values of x, the value of the gradient of the graph is numerically large, indicating that the value of the function is changing rapidly. This property was illustrated in the table and is the indicator of ill-conditioning in function evaluation.

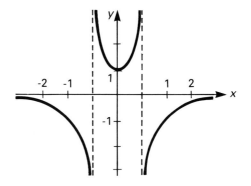

Figure 5.6

Evaluate u_2, u_3 and u_4, exactly, where $u_{n+1} = 10.1u_n - u_{n-1}$, with $u_0 = 1.001$ and $u_1 = 0.1$. Repeat the calculation carrying 3 significant figures in the working and comment on the results obtained.

Solution:

n	u_n **Exact**	u_n 3 **Sign. Figures**	**Error**
0	1.001	1.00	0.1%
1	0.1	0.1	0%
2	0.009	0.01	11%
3	−0.0091	0.001	111%
4	−0.10091	0.0001	100%

As the calculation proceeds, the error increases, showing ill-conditioning. The problem is caused by the loss of significant figures when nearly equal values are subtracted to give u_{n+1}. The problem does not arise with different values of u_0 and u_1.

The value of $f(x) = \dfrac{1}{1-x}$ is to be evaluated at $x = 0.99$. If this value of x is in error by a small amount ε, calculate the magnitude of the corresponding error in the value of $f(0.99)$.

Solution:

$$f(x) = \frac{1}{1-x}$$

$$f(0.99) = \frac{1}{0.01} = 100$$

$$f(0.99 + \varepsilon) = \frac{1}{1 - 0.99 - \varepsilon} = \frac{1}{0.01 - \varepsilon} = \frac{100}{1 - 100\varepsilon}$$

Hence $|\text{Change in } f| = |f(0.99 + \varepsilon) - f(0.90)|$

$$= \left| \frac{100}{1 - 100\varepsilon} - 100 \right|$$

$$= \left| \frac{100 - 100 + 10^4\varepsilon}{1 - 100\varepsilon} \right|$$

$$= \left| \frac{10^4\varepsilon}{1 - 100\varepsilon} \right|$$

Q Show that

$$f(x) = \frac{x^2}{\sqrt{x^2 + 1} + 1} \equiv \sqrt{x^2 + 1} - 1$$

Working to 4 significant figures, evaluate $f(x)$ for $x = 0.1$, using both formulae. Which form of $f(x)$ would you choose to use to evaluate $f(x)$ for x near to 0? Explain your answer.

Exercise 5

1. Express the following numbers to (a) 3 decimal places (b) 3 significant figures.

   (i) $\dfrac{1}{64}$ (ii) $\dfrac{1}{101}$ (iii) $2\dfrac{2}{3}$ (iv) $3\dfrac{4}{7}$

2. Express the following fixed point decimal numbers as floating point numbers with 5 significant figures.

   (i) 1201.14   (ii) 2.112   (iii) 0.00141577

3. Round the following numbers to the number of significant figures shown in brackets.

   (i)   0.01247   (3)   (ii) 28421.1   (4)
   (iii) 107.144   (2)   (iv) 91.488   (4)
   (v)   7.42999   (5)   (vi) 0.001418   (3)

4. Use your calculator to determine the value of $\sqrt{7}$. Round this value to the number of significant figures given in the table and calculate the absolute relative error in each case.

No. of sign. figures	Approx. value	Abs. rel. error
1		
5		
6		
7		
8		

5. Perform the following arithmetic operations, rounding each result to 4 significant figures.

(i) $126.4 + 0.01236$

(ii) $19.48 - 19.40$

(iii) 12.64×12.36

(iv) $1.415 \times 10^2 \div 2.736 \times 10^5$

6. Calculate, to 2 significant figures, (a) the absolute error (b) the absolute relative error and (c) the % error in each of the following approximations.

(i) $\pi \approx \dfrac{22}{7}$ (ii) $\sqrt{2} \approx 1.414$ (iii) $e \approx 2.718$

7. Solve the following equation using 3 significant figure arithmetic (i.e. round *each* calculation to 3 significant figures) to obtain the value of x.

$$x^2 + 1000x + 10 = 0$$

Evaluate $f(x) = x^2 + 1000x + 10$ for x equal to each of the roots obtained and comment on the results.

8. Solve the following quadratic equation using 3 significant figure arithmetic.

$$x^2 + 1.21x + 0.365 = 0$$

How many significant figures are needed for the calculation in the square root term in the formula to give the solutions correct to 2 significant figures?

9. Using Rules 1 and 2 from the text, perform the following calculations, giving the answers as accurately as the data will allow.

(i) $0.00041 + 2.14$

(ii) 0.005×2.91

(iii) $315.21 - 182.561$

(iv) $54.50/672.0$

(v) $27.54(18.561 - 2.72)$

(vi) $1.234(2.04 - 1.99)$

(vii) $-0.0080 - 47.35$

(viii) $89.22(34.21 - 34.19)$.

10. The formula $P = 2l + 2b$ is evaluated with $l = 1.222$ and $b = 10.5$. Give the value of P as accurately as the data will allow.

11. Evaluate $1.2345\,(123.45 - 122.33)$ justifying the number of figures quoted in the answer.

12. Obtain the values of $\sqrt{40.001}$ and $\sqrt{40.000}$ correct to 7 decimal places. How many significant figures are correct in the evaluation of $\sqrt{40.001} - \sqrt{40.000}$? Show that this expression is mathematically equivalent to

$$\frac{0.001}{\sqrt{40.001} + \sqrt{40.000}}$$

Explain why the second form of the expression is computationally better.

13. Given that

$$S = 1 + \frac{2}{3} + \frac{4}{9} + \frac{8}{27} + \ldots = 3$$

calculate the first six terms of the series correct to 4 decimal places, noting the round-off error (correct to one significant figure), in each term. Let S_n be the sum to n terms of the series and complete the following table for $n = 1, 2 \ldots 6$, to show the truncation error and the round-off error present when S_n is taken as an approximation to S.

n	S_n	$\lvert \text{Truncation Error} \rvert$	$\lvert \text{Round-off Error} \rvert$
1	1.0000	2.0000	0.0000
2			

14. Write down the maximum possible absolute error when a number is rounded to 4 decimal places. What is the maximum possible relative error when a number $a \times 10^{-b}$ is rounded to 4 significant figures?

Round the numbers   (a) $\dfrac{1}{3000}$ and (b) $\dfrac{1}{3000000}$

to   (i) 4 decimal places
and (ii) 4 significant figures.

Calculate the absolute error in (i) and the absolute relative error in (ii).

15. Show that the error in the value of

$$f(x) = x^2 - 2x - 2$$

caused by an error of ε in the value of x is ε^2 when $x = 1$.

16. The function $f(x) = x^3 + 3x^2 + 3x - 1$ has small errors of magnitude ε in the coefficients of x^2 and x. Under what conditions will the resulting error in $f(1)$ be zero? In this case, write down an expression for the absolute error in $f(2)$.

17. Show that $ax^3 + bx^2 + cx + d \equiv ((ax + b)x + c)x + d$. Count the number of multiplication operations required to evaluate each expression. Which form is computationally more efficient?
Express the polynomial $p(x) = 2x^4 + 3x^3 - x^2 + 2x - 1$ in a more efficient form for evaluation. Determine the value of $p(3.1)$.

18. Solve the equation $\qquad x^2 - ax + 0.2830 = 0$

where $\qquad$ (i) $a = 1.0640$ $\qquad$ (ii) $a = 1.0641$.

Comment on the numerical values obtained in the computation and on the condition of the equation.

19. In evaluating $a - b$, the exact value of a is 1.23446 and the exact value of b is 1.23545. The values of a and b are rounded to 4 decimal places, the value of b being rounded up. Calculate the % error in the value of $a - b$ due to this approximation. Calculate also the % error in this value if b had been rounded down to 1.2354.

20. Given that $f(x) = \tan x$, evaluate $f(x)$ for $x = 89.74$ and $x = 89.76$. Calculate the magnitude of the % change in the value of $f(x)$ when x changes from $x = 89.74$ to $x = 89.76$. Compare this with the % change in the value of x.

21. The following program generates values of x_n given by the recurrence relation $x_{n+1} = 0.6x_n^2$, with $x_0 = 5/3$.

```
10 X = 5/3
20 FOR K = 1 TO 40
30 X = 0.6 * X^2
40 PRINT X
50 NEXT
```

What are the fixed points of this iteration?
Use the theory of convergence of iterative formulae to decide if convergence of $x_{n+1} = 0.6x_n^2$ could take place to either fixed point.
Note the value given by the computer for and replace 5/3 in line 10 of the program by this value.
Run the program again and comment on the results.

22. In mathematics, $\qquad a + b = b + a$

and so $\qquad S_1 = \dfrac{1}{1} + \dfrac{1}{2} + \dfrac{1}{3} + \ldots + \dfrac{1}{32766} + \dfrac{1}{32767}$

and $\qquad S_2 = \dfrac{1}{32767} + \dfrac{1}{32766} + \ldots + \dfrac{1}{3} + \dfrac{1}{2} + \dfrac{1}{1}$

are equal. The following program calculates S_1 and S_2 and prints their values.

```
 10 S = 0
 20 FOR K = 1 TO 32767
 30 S = S + 1/K
 40 NEXT K
 50 PRINT "Value of S1 = ";S
 60 S = 0
 70 FOR K = 32767 TO 1 STEP -1
 80 S = S + 1/K
 90 NEXT K
100 PRINT "Value. of S2 =";S
```

Comment on the values obtained for S_1 and S_2 and explain the discrepancy. Modify the program to obtain corresponding results for

$$\sum_{r=1}^{32767} \frac{1}{r^2} \text{ and } \sum_{r=1}^{32767} \frac{1}{r^3}.$$

KEY POINTS

- Most numerical algorithms involve infinite processes which must be truncated, introducing a *truncation error* in the result.
- In calculations, numbers must be rounded to a fixed number of digits. The error introduced by using this approximation in place of the exact value is called a *rounding error*.
- Since data values are usually inexact and rounding takes place in calculations, care should be taken to avoid quoting the result of a calculation to more digits than can be expected to be correct.
- If the solution to a problem is extra sensitive to small changes in the data used, the problem is said to be *ill-conditioned*.

Answers to selected exercises

1 THE SOLUTION OF EQUATIONS

In-text questions

p. 4 (i) 1 root between 1 and 2 (ii) 5 roots, near to $-4.5, -1, 0, 1, 4.5$ (iii) 2 roots, between -1 and 0 and between 1 and 2

p. 6 Root lies between 0.325 and 0.32625

p. 8 1. (i) 1.5 (ii) 1.0
　　　2. 0.326

p. 10 You would expect at least 7 iterations.
　　　(i) At step 8, root is between 0.617 and 0.621, giving 0.62 correct to 2 decimal places.
　　　(ii) At step 9, root is between 1.531 and 1.533, giving 1.53 correct to 2 decimal places.

p. 21 (i) 0.74 (ii) 0.91

p. 24 (i) -0.7391　(ii) -1.8683; -0.5385; 2.5850

Exercise 1

1. 1.3
2. The root is 0.653
3. 4.5
4. (i) -0.74 (ii) 0.26　(iii) 0.36; 2.15
5. Only (ii) will converge to $x = 1$.　(i) and (iii) will converge to $x = 5$.
6. The formula would not converge.
10. 2.92
11. (i)　0.958　　(ii)　-0.193
　　(iii) 1.164　　(iv) 1.189
12. -2.70; 0.342; 2.21.
13. (i) 3.0796 (ii) 0.0556; 2.8048; 7.1550; 8.3404.
14. $x_0 = 1, 1.1, 1.2$ gives convergence to 1.124
　　$x_0 = 1.3, 1.4, 1.5$ gives convergence to 1.451
　　Negative root is -2.57
15. 16.21 units.

2 APPROXIMATING FUNCTIONS

In-text questions

p. 35 1. $1 - x^2/2! + x^4/4!$
　　　2. (i)　$1 - x + x^2 - x^3 + x^4$　(ii)　$1 - 2x + 3x^2 - 4x^3 + 5x^4$
　　　3. $p_3(x) = x + 1/6\, x^3$

p. 39 1. $p_3(x) = 0.167x^3 - 1.571x^2 + 3.935x - 2.026$
　　　　　Approximately $1.7 < x < 4.7$
　　　2. $p_3(x) = 4 - 6x + 4x^2 - x^3$ No Taylor polynomial at $x = 0$ since function is not defined for $x = 0$.

p. 42 (i) 0.125×10^{-4}　(ii) 0.125×10^{-2}

p. 48 Polynomial is $p_2(x) = 1/3\,(x - 2)(x - 4) + 1/4\,(x - 1)(x - 2)$;
　　　$p_2(2.5) = -0.0625$

Exercise 2

1. (i) $x - x^2/2 + x^3/3 - x^4/4$ (ii) $1 + 2x + 2x^2 + 4/3x^3 + 2/3x^4$ (iii) $x + x^3/3$
   (iv) $1 + x - x^2/2! - x^3/3! + x^4/4!$
2. (i) $25\ln 5 + (10\ln 5 + 5)(x - 5) + 1/2(3 + 2\ln 5)(x - 5)^2 + 1/15(x - 5)^3$
   (ii) $1/2 - \sqrt{3}/2(x - \pi/3) - 1/4(x - \pi/3)^2 - \sqrt{3}/12(x - \pi/3)^3$
   (iii) $2 + x - 8/12 - (x - 8)^2/288 + 5(x - 8)^3/20736$
   (iv) $\ln 1/2 - \sqrt{3}(x - \pi/3) - 2(x - \pi/3)^2 4\sqrt{3}/3(x - \pi/3)^3$
5. $(1 - (x - 1) + (x - 1)^2/2 - (x - 1)^3/6)e^{-1}$
6. (i) $|\text{Error}| \leqslant 1/2(\pi/4)^2 \sin \pi/4 = 0.218$ (ii) $|\text{Error}| \leqslant 1/21(0.1)^2 e^{0.1} = 0.006$
9. 18.38
10. Maximum value ≈ 3.14
11. $p_3(0) = -4.75$; $P_3(3) = 89$
12. $p_2(x) = 0.05x^2 - 0.427x + 1.153$
14. Integral $= 0.135$

3 NUMERICAL INTEGRATION

In-text questions

p. 58 1. 2.2
2. (i) 0.75 (ii) 0.708 (iii) 0.697
p. 63 $T_4 = 0.7430$ $T_8 = 0.7459$ so, to 1 decimal place, integral is 0.7
p. 67 1. 1.81
2. (i) 0.6667 (ii) 0.6369 (iii) 0.6367
Exact value, to 4 decimal places, is 0.6366
p. 74 $S_2 = 0.8828$ $S_4 = 0.8815$; Improved value is 0.8814

Exercise 3

1. 65.7
2.

	(i)	(ii)	(iii)	(iv)
Exact	1.0986	4.000	0.6366	−6.2832
Approximate	1.1167	4.2500	0.6220	−5.9569

3. (i) 0.588824; 0.552309; 0.543119; 0.5
   (ii) 4.313285; 4.358037; 4.369456; 4.4
4. Repeating 1: 65.0
   ,,   2: (i) 1.1000 (ii) 4.0000 (iii) 0.6369 (iv) −6.2975
   ,,   3; (i) 0.541401; 0.540137; 0.540056; 0.5401
              (ii) 4.366236; 4.372954; 4.373262; 4.373
5. Errors have magnitude (i) 3.6×10^{-7} and (ii) 6.3×10^{-7}
6. $S_4 = 0.72056$; $S_8 = 0.72067$; 0.72068
7. $T_2 = 0.3760$; $|\varepsilon| \leqslant 0.01$ $S_2 = 0.3858$; $|\varepsilon| \leqslant 0.002$
8. $T_2 = 0.030$; $S_2 = 0.032$
   No upper limit can be calculated for the magnitude of the truncation error.
   Actual errors are 0.005 and 0.003.

4 THE SOLUTION OF DIFFERENTIAL EQUATIONS

In-text questions

p. 82 Using forward differences,

h	0.1	0.05	0.01
f'(1.5)	−0.417	−0.430	−0.442
|error|	0.027	0.014	0.002

Using central differences,

h	0.1	0.05	0.01
f'(1.5)	−0.446	−0.445	−0.4445
|error|	0.002	0.001	0.0001

p. 85

t	Exact	Approximate
20	39.207947	39.207555
40	38.431578	38.430808
60	37.670581	37.669451
80	36.924654	36.923176
100	36.193497	36.191686

p. 88 (i)

x	0.01	0.02	0.03	0.04
y	1.010	1.020	1.030	1.041

p. 96 (ii)

x	0.1	0.2	0.3	0.4	0.5
y	1.000	1.010	1.029	1.056	1.090

Exercise 4

1. Approximations to f'(0) are
   Forward formula: −0.0997;   −0.0200;   −0.0040
   Central formula: 0   ;   0   ;   0
   Approximations to f'(1.5) are
   Forward formula: −0.9979;   −0.9986;   −0.9978
   Central formula: −0.9908;   −0.9972;   −0.9975
2. Both formulae give   f'(0.5) = −0.4
   Using forward differences   f"(0.5) = 1.6
   Using central differences   f"(0.5) = 0

3. $f''(x_0) = 1/h^2\{f(x_0 + 2h) - 2f(x_0 + h) - f(x_0)\}$

4. (i) 1.06 (ii) 1.55 (iii) 4.21 (iv) −0.18 (v) 0.466 (vi) 1.16
5. 1.1159
6. $y(0.1) = 2.2150$; $y(0.2) = 2.4630$; $y(0.3) = 2.7473$; $y(0.4) = 3.0715$;
$y(0.5) = 3.4394$
7. Calculated values   1.80975   3.29198   4.50596
    Exact values   1.81269   3.29680   4.51188

5 ERRORS IN NUMERICAL PROCESSES

In-text questions

p. 103 0.09983; 2 terms; $x^2/6$
p. 106 1. 2.150×10^4
       2. (i) 0.0140 3 significant figures (ii) 72000 4 significant figures
p. 107 1. (i) 203.7 with errors 0.036; 1.77×10^{-4}
          (ii) 203.66 with errors 0.004; 1.96×10^{-5}
       2. 5×10^{-4}; 5×10^{-4} (i) 0.0033   Absolute error = 0.000033
          (ii) 0.003333 Absolute relative error = 0.0001
p. 113 Formula is ill-conditioned for x near to 1.1, but not near to $x = 6.1$
p. 115 0.004988; 0.005

Exercise 5

1. (i) 0.016; 0.0156   (ii) 0.010; 0.00990
   (iii) 2.667; 2.67   (iv) 3.571; 3.57
2. 1.2011×10^3 (ii) 2.1120×10^0 (iii) 1.4158×10^{-3}
3. (i) 0.0125 (ii) 28420 (iii) 110
   (iv) 91.49 (v) 7.4300 (vi) 0.00142
5. (i) 126.4   (ii) 0.08 (significant figures have been lost)
   (iii) 156.2   (iv) 5.172×10^{-4}.
6. Results will depend on the calculator used.
7. 1000, 0.
8. Equal roots −0.605; 4 figures should be carried in the square root calculation.
9. (i) 2.14 (2 decimal places, using Rule 1)
   (ii) 0.01 (1 significant figures, using Rule 2)
   (iii) 132.65 (2 decimal places, using Rule 1)
   (iv) 0.08110 = 8.110×10^{-2} (using Rule 2)
   (v) $27.54 \times 15.84 = 436.2$ (vi) $1.234 \times 0.005 = 0.06$
   (vii) −47.36   (viii) $89.22 \times 0.02 = 2$
10. 23.4
11. 1.38
12. 3 significant figures
14. (a) (i) Absolute error 0.000033   (ii) Relative Error 1×10^{-4}
     (b) (i) Absolute error $(3000000)^{-1}$ (ii) Relative Error 1×10^{-4}
16. Errors in coefficients have opposite signs; 2ε
18. (i) 0.5271, 0.5369 (ii) 0.5232, 0.5408 Equation is ill-conditioned
19. 1%; 9%
20. % change in $f(x) = 9\%$; % change in $x = 0.02\%$
21. Fixed points are 0 and 5/3.

British Library Cataloguing in Publication Data

West, Elizabeth
    Numerical Analysis. – (MEI Structured
    Mathematics Series)
    I. Title II. Series
    515

ISBN 0 340 57303 1

First published 1993
Impression number 10 9 8 7 6 5 4 3 2 1
Year   1998 1997 1996 1995 1994 1993

Typeset by Keyset Composition, Colchester, Essex
Printed in Great Britain for the education publishing division of
Hodder & Stoughton Ltd, Mill Road, Dunton Green, Sevenoaks,
Kent TN13 2YA by Thomson Litho Ltd.